New Antiepileptic Drugs

Acknowledgment
The publication of this volume has been supported by a grant from Lepetit Group, Wellcome and Ciba-Geigy.

New Antiepileptic Drugs

EDITED BY

FRANCESCO PISANI

Institute of Neurology, University of Messina, Messina, Italy

EMILIO PERUCCA

Clinical Pharmacology Unit, Department of Internal Medicine and Therapeutics,
University of Pavia, Pavia, Italy

GIULIANO AVANZINI

National Neurological Institute C. Besta, Milan, Italy

ALAN RICHENS

Department of Pharmacology and Therapeutics, University of Wales College of Medicine,
Cardiff, UK

EPILEPSY RESEARCH

SUPPLEMENT NO. 3

1991

ELSEVIER
AMSTERDAM · NEW YORK · OXFORD

ISBN 0–444–81392–6
ISSN 0920–1211 (series)

Published by:

Elsevier Science Publishers B.V. (Biomedical Division)
P.O. Box 211, 1000 AE Amsterdam, The Netherlands.

Sole distributors for the USA and Canada:
Elsevier Science Publishing Company, Inc.
52 Vanderbilt Avenue, New York, NY 10017, USA.

Library of Congress Cataloging-in-publication Data

New antiepileptic drugs / edited by Francesco Pisani ... [et al.].
      p. cm. -- (Epilepsy research. Supplement : no. 3)
   Includes bibliographical references and index.
   ISBN 0-444-81392-6
   1. Epilepsy--Chemotherapy.   2. Anticonvulsants--Testing.
  I. Pisani, Francesco.   II. Series.
   [DNLM: 1. Anticonvulsants--therapeutic use.   2. Epilespsy-drug
therapy.   W1 EP455KEA no. 3 / QV 85 N532]
   RC374.C48N49   1991
   616.8′5306--dc20
   DNLM/DLC
   for Library of Congress   91-4270
                                            CIP

Printed in the Netherlands

List of contributors

B. ALDENKAMP
Instituut voor Epilepsiebestrijding, Achterweg 5, 2103 SW Heemstede, The Netherlands

H. ANHUT
Gödecke AG Research Institute, Mooswaldallee 1–9, 7800 Freiburg, Germany

C. ARTESI
First Neurological Clinic, University of Messina, 98013 Contesse-Messina, Italy

G. AVANZINI
National Neurological Institute C. Besta, Via Celona 11, 20133 Milan, Italy

L. BERTILSSON
Department of Clinical Pharmacology, Karolinska Institute, Huddinge University Hospital, 14186 Huddinge, Sweden

D. BETTUCCI
Department of Neurology, Turin University School of Medicine at Novara, Ospedale Maggiore, 28100 Novara, Italy

F. DE BEUKELAAR
Janssen Research Foundation, Turnhoutseweg 30, 2340 Beerse, Belgium

A.M. BEUN
Instituut voor Epilepsiebestrijding, Achterweg 5, 2103 SW Heemstede, The Netherlands

R. CANTELLO
Department of Neurology, Turin University School of Medicine at Novara, Ospedale Maggiore, 28100 Novara, Italy

D. CHADWICK
Department of Neurology, Walton Hospital, Liverpool L9 1AE, UK

A. CHAPMAN
Department of Neurology, Institute of Psychiatry, De Crespigny Park, London SE5 8AF, UK

S.S. CHATTERJEE
Dr. Willmar Schwabe Arzneimittel, Willmar Schwabe Strasse 4, 7500 Karlsruhe 41, Germany

R. DUANE SOFIA
Preclinical and Clinical Research and Development, Wallace Laboratories, Cranbury, NJ 08512, USA

M. ENGELSMAN
Instituut voor Epilepsiebestrijding, Achterweg 5, 2103 SW Heemstede, The Netherlands

A. FAZIO
First Neurological Clinic, University of Messina, 98013 Contesse-Messina, Italy

F. FEDERICO
Institute of Neurology, School of Medicine, University of Bari, 70100 Bari, Italy

M. FOOT
Clinical Research Division, Warner Lambert Co, Eastleigh, Hants SO5 3ZQ, UK

E. FORCEVILLE
Instituut voor Epilepsiebestrijding, Achterweg 5, 2103 SW Heemstede, The Netherlands

M. GIANELLI
Department of Neurology, Turin University School of Medicine at Novara, Ospedale Maggiore, 28100 Novara, Italy

L. GRAM
Dianalund Epilepsy Hospital, 4293 Dianalund, Denmark

L.E. GUSTAVSON
Abbott Laboratories, Abbott Park, IL 60064, USA

A. VON HODENBERG
Gödecke AG Research Institute, Mooswaldallee 1–9, 7800 Freiburg, Germany

T. ITO
Research Laboratories, Dainippon Pharmaceutical Co. Ltd, Enoki 33–94, Suita/Osaka 564, Japan

D. JANZ
Department of Neurology, Klinikum Charlottenburg, Free University of Berlin, Spandauer Damm 130, 1000 Berlin 19, Germany

K. KLESSING
Dr. Willmar Schwabe Arzneimittel, Willmar Schwabe Strasse 4, 7500 Karlsruhe 41, Germany

P. KLOSTERSKOV JENSEN
R & D Department, Ciba-Geigy Ltd, 4002 Basel, Switzerland

L. KRAMER
Preclinical and Clinical Research and Development, Wallace Laboratories, Cranbury, NJ 08512, USA

P.J. LEWIS
Merrell Dow Research Institute, Winnersh Berkshire RG11 5HQ, UK

T. MANT
Guy's Drug Research Unit, Newcomer Street 6, London SE1, UK

C. MARESCAUX
Clinique Neurologique C.H.U., Place de l'Hôpital, Strasbourg Cedex 67091, France

J.F. MCKELVY
Abbott Laboratories, Abbott Park, IL 60064, USA

H. MEINARDI
Instituut voor Epilepsiebestrijding, Achterweg 5, 2103 SW Heemstede, The Netherlands

B. MELDRUM
Department of Neurology, Institute of Psychiatry, De Crespigny Park, London SE5 8AF, UK

H.B. MENGEL
Novo Nordisk A/S, 2880 Bagsvaerd, Denmark

R. MICHELUCCI
Neurological Clinic, University of Bologna School of Medicine, 40123 Bologna, Italy

H. MIYAZAKI
Research Laboratories, Dainippon Pharmaceutical Co. Ltd, Enoki 33–94, Suita/Osaka 564, Japan

F. MONACO
Neurological Clinic, Sassari University Medical School, 07100 Sassari, Italy

K.T. MUIR
Fisons Pharmaceuticals, Divisional Research and Development, Jefferson Road, Rochester NY 14603, USA

J.P. MUMFORD
Merrell Dow Research Institute, Winnersh Berkshire RG11 5HQ, UK

R. MUTANI
Department of Neurology, Turin University School of Medicine at Novara, Ospedale Maggiore, 28100 Novara, Italy

B. NUNES
Instituut voor Epilepsiebestrijding, Achterweg 5, 2103 SW Heemstede, The Netherlands

G. OTERI
First Neurological Clinic, University of Messina, 98013 Contesse-Messina, Italy

G.C. PALMER
Fisons Pharmaceuticals, Divisional Research and Development, Jefferson Road, Rochester, NY 14603, USA

J.L. PERHACH
Preclinical and Clinical Research and Development, Wallace Laboratories, Cranbury, NJ 08512, USA

R. DI PERRI
First Neurological Clinic, University of Messina, 98013 Contesse-Messina, Italy

E. PERUCCA
Clinical Pharmacology Unit, Department of Internal Medicine and Therapeutics, University of Pavia, Piazza Botta 10, 27100 Pavia, Italy

M.W. PIERCE
Abbott Laboratories, Abbott Park, IL 60064, USA

F. PISANI
First Neurological Clinic, University of Messina, 98013 Contesse-Messina, Italy

E.H. REYNOLDS
Department of Neurology, King's College Hospital, Denmark Hill, London SE5 9RS, UK

A. RICHENS
Department of Pharmacology and Therapeutics, University of Wales College of Medicine, Heath Park, Cardiff CF4 4XN, UK

A. ROSENBERG
Preclinical and Clinical Research and Development, Wallace Laboratories, Cranbury, NJ 08512, USA

M. RUSSO
First Neurological Clinic, University of Messina, 98013 Contesse-Messina, Italy

A. DE SARRO
Institute of Pharmacology, School of Medicine, University of Messina, 98100 Messina, Italy

G. DE SARRO
Institute of Pharmacology, School of Medicine, University of Messina, 98100 Messina, Italy

G. SATZINGER
Gödecke AG Research Institute, Mooswaldallee 1–9, 7800 Freiburg, Germany

D. SCHMIDT
Department of Neurology, Free University of Berlin, Spandauer Damm 130, 1000 Berlin 19, Germany

M. SCHMUTZ
R & D Department, Ciba-Geigy Ltd, 4002 Basel, Switzerland

G. SECHI
Neurological Clinic, Sassari University Medical School, 07100 Sassari, Italy

M. SEINO
National Epilepsy Center, Shizuoka Higashi Hospital, Shizuoka, Japan

U. STEIN
Dr. Willmar Schwabe Arzneimittel, Willmar Schwabe Strasse 4, 7500 Karlsruhe 41, Germany

P.D. SUZDAK
Novo Nordisk A/S, 2880 Bagsvaerd, Denmark

C.A. TASSINARI
Neurological Clinic, University of Bologna School of Medicine, 40123 Bologna, Italy

T. TOMSON
Department of Neurology, Söder Hospital, 10064 Stockholm, Sweden

G.R. TRIMARCHI
Institute of Pharmacology, School of Medicine, University of Messina, 98100 Messina, Italy

R. TRIO
First Neurological Clinic, University of Messina, 98013 Contesse-Messina, Italy

x

L. TRITSMANS
Janssen Research Foundation, Turnhoutseweg 30, 2340 Beerse, Belgium

J.C. VINCENT
Biocodex Research Laboratories, 19 Rue Barbes, 92120 Montrouge, France

J. WALLACE
Clinical Research Division, Parke-Davis, 2800 Plymouth Road, Ann Arbor, MI 48105, USA

K.D. WOLTER
Ciba-Geigy Pharma Division, 566 Morris Avenue, Summit, NJ 07901, USA

A.W.C. YUEN
Wellcome Research Laboratories, Langley Court, South Eden Park Road, Beckenham, Kent BR3 3BS, UK

List of contents

Preface

The main aim of this volume is to offer an up-to-date review of the most recent advances in antiepileptic drug development, considered from various points of view: (i) general, by taking into account the size of refractory epilepsy and its related problems; (ii) experimental, by exploring the mechanisms of epileptogenesis and the possibility of influencing it through drugs; (iii) clinical, by describing the results obtained with compounds currently at an advanced stage of testing. The pharmacological fight against epilepsy started many centuries ago, when Hippocrates realised that epilepsy has a natural cause and, consequently, must be treated with a natural (and not a supernatural!) remedy. Even though since that era science progressed, the challenge to find remedies against epilepsy still continues. We hope that this volume will provide an additional stimulus to meet this challenge, in the hope to witness shortly the final victory.

F. Pisani, E. Perucca
G. Avanzini, A. Richens

New Antiepileptic Drugs (Epilepsy Res. Suppl. 3)
Editors: F. Pisani, E. Perucca, G. Avanzini, A. Richens
© 1991 Elsevier Science Publishers B.V. (Biomedical Division)

Introduction

Raoul di Perri[1] and Dieter Janz[2]

[1]*First Neurological Clinic of the University of Messina, Italy, and* [2]*Department of Neurology, Klinikum Charlottenburg, Free University of Berlin, Germany*

Since the introduction of monitoring of antiepileptic drugs, namely about three decades ago, many goals have been achieved in the treatment of epilepsy. Aspects such as patient compliance, drug kinetics and interactions, drug toxicity and efficacy, real and false drug resistance, mono- and polypharmacy have been well recognized and evaluated. As a direct consequence, the overall prognosis for epilepsy has improved noticeably and it is now established that seizure remission occurs in at least 75% of patients following adequate treatment with one of the currently available antiepileptic drugs. Apart from a very limited number of patients who are suitable for surgery, the remainder, i.e. about 25%, still continue to have frequent seizures. The life of these patients is crippled not only by the recurrence of seizures, but also by the development of drug toxicity which occurs especially with polypharmacy and with high doses of drugs.

The modern history of the development of antiepileptic drugs started in the middle of the last century, when Sir Charles Locock introduced potassium bromide in the treatment of catamenial epilepsy. He made this attempt on the basis of the known ability of bromide salts to produce sedative effects and impotence. In 1912, Hauptmann reported on the strong antiepileptic action of phenobarbitone in man; also in this case the drug was tested in human epilepsy on the basis of its strong sedative effects. Within a few years phenobarbitone became the drug of choice in the treatment of epilepsy.

This early era, characterized by the belief that anticonvulsant and sedative effects of drugs were in some way linked, ended when the increasing availability of models of experimental epilepsy made possible a different approach in identifying new drugs. In 1937, Merritt and Putnam tested several compounds in electrically induced seizures in cat and, among these compounds, phenytoin proved to be one of the most potent[1]. When phenytoin was found to be very effective also in human epilepsy without exerting sedative effects, two new concepts were established: (i) antiepileptic and sedative effects can occur separately, and (ii) the anticonvulsant action in animals may predict efficacy in man. Hence, testing substances in the laboratory became a valid, less costly and irreplaceable tool to identify and develop new drugs for human epilepsy. In this period several compounds were identified and marketed as antiepileptic drugs: trimethadione, mephenytoin, paramethadione, primidone, methsuximide, ethosuximide.

After 1960 (the year of the introduction of ethosuximide), excepting some benzodiazepines approved as adjunctive treatment, only two major drugs, namely carbamazepine and valproic acid, have become available for widespread use

4

in the therapy of epilepsy. This period was dominated by research on clinical pharmacology of the already existing antiepileptic drugs and by the hope that, by using these drugs more rationally, epilepsy could be adequately controlled. The practice of antiepileptic drug monitoring has resulted in a remarkable therapeutic improvement with a consequent increase in the number of seizure-free patients. However, in spite of these undeniable advances, the awareness of the inadequacy of the currently available antiepileptic armamentarium with respect to both safety and efficacy has become increasingly clear in recent years.

In the meantime, the improved understanding of the basic mechanisms of epileptogenesis has led to a new approach in the development of new antiepileptic drugs. This approach is characterized by an effort to design new molecules acting in order to interfere with specific mechanisms known to be involved in epileptogenic or epileptic phenomena. The development of drugs which modify GABA activity or inhibit excitatory aminoacids is an expression of this scientific direction.

The present book deals with the most recent developments in this field. The first section is concerned with general aspects, such as the possible presence of cognitive deterioration in patients with refractory epilepsy and the influence of antiepileptic drugs on the natural history of epilepsy. The following section is dedicated to experimental aspects concerning the basic mechanisms through which a compound may exert antiepileptic action. Pharmacokinetics, drug interactions and clinical data of the new antiepileptic drugs are given in a separate section. A number of new drugs are described with regard to their pharmacological and clinical features, and the results of recent clinical trials are reported.

Of course, this publication is not intended to compete with reference books on antiepileptic drug therapy (2–4), but it is simply meant to provide an updated review on the present state of research on new antiepileptic drug development and to stimulate further interest in this field.

References

1   Friedlander, W.J. (1986) Epilepsia 27(3): 1–21.
2   Frey, H.H., Janz, D. (1985) Antiepileptic drugs. Handbook of experimental pharmacology. Springer, Berlin, Heidelberg, New York, Tokyo.
3   Meldrum, B.S., Porter, R.J. (1986) New anticonvulsant drugs. John Libbey, London, Paris.
4   Levy, R.H., Dreifuss, F.E., Mattson, R.H., Meldrum, B.S., Penry, J.K., Hessie, B.J. (1989) Antiepileptic drugs. Raven Press, New York.

General Aspects

New Antiepileptic Drugs (Epilepsy Res. Suppl. 3)
Editors: F. Pisani, E. Perucca, G. Avanzini, A. Richens
© 1991 Elsevier Science Publishers B.V. (Biomedical Division)

CHAPTER 1

Mental deterioration in a population with intractable epilepsy

Harry Meinardi, Anne Marie Beun, Bert Aldenkamp, Belina Nunes, Martijn
Engelsman, Etienne Forceville

Instituut voor Epilepsiebestrijding, Achterweg 5, 2103 SW Heemstede, The Netherlands

Introduction

In a recent monograph on 'Intractable Epilepsy' no references to dementia or cognitive deterioration will be found in the index[1]. Yet the notion that epilepsy is associated with dementia is ubiquitous in the literature, even in the pre-scientific era, usually considered to end at the time of Hughlings Jackson, as in the post-Jackson and post-Hans Berger writings. Several explanations are given for the possible presence of cognitive deterioration:

1. Gowers[2] considered dementia to be a sequel to seizures.
2. Jelgersma[3] hypothesized the accumulation of toxic substance(s) in the body and the brain responsible both for the intermittent occurrence of seizures and chronically for a continuous dementia.
3. Lennox[4] was more discriminative and considered five factors to be at play either alone or in combination, i.e. genes, brain damage, psychological handicaps, the seizure process and antiepileptic drugs.

Nearly half a century later the epidemiology of dementia in epilepsy and the role of Lennox's five factors are still unclear. There are a number of obstacles that are difficult to circumvent. Many seizures may also imply extensive brain damage and are also more likely to be accompanied by high doses of antiepileptic drugs.

Definition of dementia

Dealing with dementia the problem of definition arises. Although the exact definition is controversial, most authors appear to agree that dementia is a clinical syndrome of deterioration of mental functioning revealing – in the presence of intact consciousness – defects in three or more of the following aspects: memory, visuo-spatial ability, personality, language and other acquired abilities like arithmetic, reading, writing, judgment and abstraction capacity. Often it is added that the acquired deterioration has to interfere with social and/or professional performance. Within the context of this definition the dementia may be stationary, progressive or even reversible.

An important point of debate is whether there is a true or only apparent difference between people who show arrest of development of mental functioning and those who deteriorate.

In the field of epileptology generally the stationary and progressive types of dementia are the main object of study and discussion. A number of studies, particularly in children, are also concerned with the arrest of development of mental functioning. For example in studies by Bourgeois et al.[5] and by Rodin[6] it was noticed on follow up that certain children with epilepsy developed slower than their schoolmates and thus presented with a lower, seemingly declining, IQ. Both these authors remarked that there was a clear association between high phenobarbitone levels and lack of intellectual development. Dementia due to intoxication is often a reversible dementia. Zaret and Cohen[7] reported an exceptional valproic acid-induced dementia in a 21 year old man with epilepsy. Recent studies on the influence of transient cognitive impairment on performance of children at school might be classified as studies on reversible dementia[8,9].

Lesser et al.[10] presented an extensive review of the literature on mental deterioration in epilepsy. They summarized their paper as follows: 'A variety of factors could potentially influence the occurrence of mental deterioration in epilepsy, including seizure type, age of seizure onset, seizure duration and seizure severity. The available literature suggests that measures of severity are more predictive of progressive decreases in intellectual functioning. There is also evidence suggesting that seizure severity and cognitive deterioration might both be the result of underlying pathophysiologic abnormalities in some cases. In the majority of patients with epilepsy, however, with relatively less severe disease, there is little evidence for cognitive deterioration. Total seizure number also has an inverse correlation with level of psychosocial functioning in some studies, whereas others have found that patients with emotional difficulties have fewer seizures. In the case of emotional deterioration, the impact of interpersonal relationships and other environmental factors upon psychosocial outcome seems clear and the evidence for specific pathophysiologic explanations for emotional deterioration, less convincing.'

Apparently the theories of Kretschmer[11], who hypothesized that, even though epilepsy occurred among several somatotypical constitutions, in particular athletic and dysplastic somato-types appeared to be epilepsy prone, have not been further pursued. Consequence of the Kretschmer hypothesis would be that people with a viscous (glyschroid) personality would be predisposed to become epileptic and hence the characteristics of epileptic dementia would be simply the expression of their constitution and would be encountered particularly in athletic or dysplastic people with epilepsy.

Dementia in a population resident in a special centre for epilepsy

The percentage of people with epilepsy who suffer from mental deterioration is probably small. It is also likely that they will be found in the category that needs special care. In an attempt to clarify this question the people with epilepsy receiving residential care at the special centre for epilepsy 'Instituut voor Epilepsiebestrijding' in the Netherlands were screened for evidence of mental deterioration. Amongst 520 patients present at the onset of the study, 300 met the requirement that at least three psychological assessments had to be available.

Criteria for the diagnosis of mental deterioration were: a fall in scores on standardized intelligence tests of at least one standard deviation (i.e. > or = 15 points) or a loss in mental age > or = 18 months. Twenty patients met these criteria (group D = dementia). A control group of twenty patients matched for age, sex and level of intellectual functioning was formed from the same population (group S = stable). The patient histories were examined for type and number of seizures, type, number and dosages of antiepilep-

TABLE 1

BASIC DETAILS OF THE PATIENT GROUPS

	GROUP D		GROUP S	
Total (male/female)	20	(10/10)	20	(10/10)
Mean actual age (range in years)	45.9	(26–68)	40.2	(26–57)
Initial mean IQ (range)	70 [n=18]	(47–87)	58.7 [n=17]	(40–84)
Recent mean IQ (range)	46.6 [n=18]	(26–60)	57.5 [n=17]	(42–80)
Initial mental age (years)	5.05 [n=2]		2.69 [n=3]	
Recent mental age (years)	3.04 [n=2]		3.05 [n=3]	
Mean age at initial IQ test (range in years)	32	(8–55)	29.2	(14–50)
Mean age at recent IQ test (range in years)	43	(24–65)	38.3	(23–57)

TABLE 2

NUMBER OF SEIZURES

	GROUP D		GROUP S	
Year	1977	1987	1977	1987
Mean (range)	65.4 (0–292)	68.3 (0–271)	63.25 (0–250)	51.8 (0–128)

TABLE 3

NUMBER OF SEIZURE TYPES

	GROUP D		GROUP S	
number of seizure types/year	1977	1987	1977	1987
	N (%)	N (%)	N (%)	N (%)
0	1 (5)	1 (5)	2 (10)	3 (15)
1–2	3 (15)	7 (35)	1 (5)	7 (35)
3 or more	16 (80)	12 (60)	17 (85)	10 (50)

TABLE 4

NUMBER OF PATIENTS/SEIZURE TYPE

	GROUP D		GROUP S	
seizure type	1977	1987	1977	1987
tonic-clonic	15	16	15	12
tonic	19	16	16	12
complex partial	11	11	12	8
atonic[*]	8	4	6	3
myoclonic	8	5	8	6
absences	13	4	10	5

* In accordance with the international classification of seizures a number of 'atonic' seizures have been reclassified as tonic seizures in 1987.

10

TABLE 5

NUMBER OF SEIZURES/SEIZURE TYPE

	GROUP D		GROUP S	
seizure type	1977	1987	1977	1987
tonic clonic	216	424	265	173
tonic	332	634	486	463
complex partial	226	259	115	165
atonic*	183	8	67	31
myoclonic	52	23	245	130
absences	299	18	87	74

*In accordance with the international classification of seizures a number of 'atonic' seizures have been reclassified as tonic seizures in 1987.

TABLE 6

NUMBER OF ANTIEPILEPTIC DRUGS/PATIENT

	GROUP D		GROUP S	
number of AED*	1977 N (%)	1987 N (%)	1977 N (%)	1987 N (%)
0 - 2	4 (20)	8 (40)	1 (5)	11 (55)
3	10 (50)	10 (50)	11 (55)	7 (35)
4 - 5	6 (30)	2 (10)	8 (40)	2 (10)

* = antiepileptic drugs.

tic drugs, and minor head traumas in two years ten years apart (1977 and 1987). Furthermore, information from CT-scan and MRI was examined if available.

Psychological tests. The findings of psychological tests in the two patient groups are shown in Table 1. On the average 4.8 tests (range 3–11) were available in group D as compared to 3.6 (range 2–13) in group S. The period of deterioration in group D ranged from 2 to 23 years (mean 10.6 years) with a mean drop in IQ of 23.4 points (range 15–39).

Seizure variables. Different aspects of the seizure pattern in the two groups were looked into. For this purpose all data from 1977 and 1987 were collected (Tables 2–5).

Medication. The average number of antiepileptic drugs taken by the patients in the two groups during the criterion years was investigated. Also the average dosage of each drug and the number of patients on each particular drug were compared. The data are shown in Tables 6–8.

Head injury. One of the factors supposed to contribute to mental deterioration in people with epilepsy is head trauma. Similarly head injury is supposed to contribute to neuropsychological deficits in alcoholism[12] and boxing[13]. Since 1977 a register is kept at the 'Instituut voor Epilepsiebestrijding' of head injuries sustained by the patients who required medical attention. However,

TABLE 7

TYPE OF ANTIEPILEPTIC DRUG USED

	GROUP D		GROUP S	
antiepileptic drug	1977 N	1987 N	1977 N	1987 N
carbamazepine	20	19	18	17
valproate	20	16	18	17
phenytoin	13	7	14	7
phenobarbitone	2	4	5	5
ethosuximide	1	0	6	1
primidone	1	1	0	0
clobazam	3	0	2	0
sulthiame	2	0	2	0
acetazolamide	1	1	1	0

TABLE 8

MEAN DOSE OF ANTIEPILEPTIC DRUG (CALCULATED/ALL PATIENTS ON THAT DRUG)

	GROUP D		GROUP S	
antiepileptic drug	1977 mg/pat	1987 mg/pat	1977 mg/pat	1987 mg/pat
carbamazepine	1215	1147	1133	1164
valproate	1965	1819	1967	1782
phenytoin	272.7	313.4	262.5	261
phenobarbitone	125	131.2	140	105
ethosuximide	750	0	1350	750
primidone	750	750	0	0

there is no book keeping about the severity in this register. In the period 1977–1987 the Dementia Group suffered an average of 10.2 head injuries while this number was 5.2 or the Stable Group.

Radiological examinations. Only information concerning CT-scans and MRI was analyzed. Most of the patients in both groups had one or more CT-scans. The data of the most recent scan are shown in Table 9. The number of patients in the groups from whom no scans were available is also given.

Discussion

The data suffer from the handicap of retrospective studies in that relevant data like psychological testing, serum levels of antiepileptic drugs, electroencephalographic and radiological examinations were not produced simultaneously. In fact, the number of serum level determinations and EEGs differed too much in the two groups to permit inclusion in the analysis. Nevertheless the two groups were clearly different with respect to a number of important measures. The average

TABLE 9

RADIOLOGICAL EXAMINATIONS

	GROUP D	GROUP S
CT-scan	N	N
not done	1	2
normal	3	3
cerebral atrophy	0	1
cerebellar atrophy	4	5
cerebro-cerebellar atrophy	3	2
vascular lesions	1	1
vasc. lesions and cerebral atrophy	2	1
vasc. lesions, cer. atrop. and hydroc.	1	0
calcifications and vascular lesions	0	1
cerebral hypoplasia	3	0
hydrocephalus	1	1
porencephalus	0	1
cerebral hypopl. and cerebellar atrophy	0	1
cerebellar atrophy and porencephalus	0	1
post-ictal oedema	1	0
MRI	N	N
not done	18	20
normal	1	–
leuco-encephalopathy	1	–

number of seizures increased in Group D (1977 = 65.4; 1987 = 68.3) and decreased in Group S (1977 = 63.25; 1987 = 51.8; Table 2). The number of tonic-clonic and tonic seizures also greatly increased in Group D (1977 tonic-clonic 216, tonic 332; 1987 tonic-clonic 424, tonic 634) while in Group S the number of tonic seizures remained stable (1977 = 486; 1987 = 463) and the number of tonic-clonic seizures declined (1977 = 265; 1987 = 173; Table 5). In this respect our findings are different from those of Thompson et al.[14].

Trimble[15], in association with several authors, has reviewed the literature on the adverse effects of antiepileptic drugs on cognitive functioning and has contributed further information to this topic. Yet no major differences could be detected amongst the patients in Groups D and S with re-

spect to type of drug (Table 7) or mean dosage of the drugs (Table 8), though the general trend to reduce polytherapy was more evident in Group S than in Group D (Table 6). Contrary to Thompson et al.[14] no evidence was found of a higher number of episodes of clinical and biochemical intoxication in Group D. However, this may have been due to methodological differences in data collection. In concordance with the Chalfont data[14], differences were present with respect to the number of head injuries, a higher number being sustained by Group D than by Group S. However, as mentioned above, Group D had also more frequent generalized tonic-clonic and tonic seizures. The outcome of the radiological examinations does not reveal important differences or at least the numbers are too small to draw conclusions about the significance of the findings.

As a spin-off of this investigation the records of intelligence test data of 300 patients with severe epilepsy and subnormal intelligence have been studied. This offered an opportunity to compare the sub-test profiles with those of persons suffering from mental retardation without epilepsy (Forceville et al, personal communication). The patients with epilepsy resembled other samples of people with epilepsy with normal or higher IQ's in that the verbal scores were generally higher than the performance scores and there was a relative difficulty with Coding and Digit Span. In these respects these patients differed from the control group of mentally retarded patients without epilepsy. Further studies are in progress to compare patients with epilepsy and patients deteriorated after head injury without epilepsy.

References

1   Schmidt, D., Morselli, P.L. (eds) (1986) Intractable Epilepsy. Raven Press, New York.

2   Gowers, W.R. (1885) Epilepsy and Other Chronic Convulsive Diseases: Their Causes, Symptoms and Treatment. William Wood, London.

3   Jelgersma, G. (1926)Leerboek der Psychiatrie, deel III, 3e druk.

4   Lennox, W.G. (1942) Am. J. Psych. 99: 174–180.

5   Bourgeois, B.F.D., Prensky, A.L., Palkes, H.S., Talent, B.K. Busch, S.G. (1983) Ann Neurol 14: 438–444.

6   Rodin, E.A., Schmaltz, S., Twitty, G. (1986) Dev. Med. Child Neurol. 28: 25–33.

7   Zaret, B.S., Cohen, R.A. (1986) Epilepsia 27: 234–240.

8   Kasteleijn-Nolst Trenité, D.G.A., Bakker, D.J., Binnie, C.D., Buerman. A., and van Raaij, M. (1988) Epilepsy Res. 2: 111–116.

9   Siebelink, B.M., Bakker, D.J., Binnie, C.D., Kasteleijn-Nolst Trenité, D.G.A. (1988) Epilepsy Res. 2: 117–121.

10 Lesser, R.P., Lüders, H., Wyllie, E., Dinner, D.S., Morris III, H.H. (1986) Epilepsia 27 (suppl.2): S105–S123.

11 Kretschmer, E. (1951) Körperbau und Character. Springer Verlag, Berlin, Göttingen, Heidelberg.

12 Hillbom, M. and Holm, L. (1986) J. Neurol. Neurosurg. Psychiatry 49: 1348–1353.

13 Grewel, F. (1941) Ned. T. Geneesk. 85: 154–160.

14 Thompson, P.J., Sander, J.W.A.S., Oxley, J. (1987) In: P. Wolf, M. Dam, D. Janz, F.E. Dreifuss (eds.), Advances in Epileptology XVI, Raven Press, New York, pp. 611–614.

15 Trimble, M.R., Reynolds, E.H. (eds) (1988) Epilepsy, Behaviour and Cognitive Function. John Wiley & Sons, Chichester.

New Antiepileptic Drugs (Epilepsy Res. Suppl. 3)
Editors: F. Pisani, E. Perucca, G. Avanzini, A. Richens
© 1991 Elsevier Science Publishers B.V. (Biomedical Division)

CHAPTER 2

The influence of antiepileptic drugs on the natural history of epilepsy

E.H. Reynolds

Department of Neurology, King's College Hospital, Denmark Hill, London SE5 9RS, UK.

Changing view of epilepsy prognosis

Whether antiepileptic drugs influence the natural history of epilepsy is an important question which has received little attention partly because, for understandable reasons, we remain ignorant of the natural history of untreated epilepsy[1,2,3]. When Rodin[4] reviewed the literature on the prognosis of epilepsy for approximately a century prior to 1968 he concluded that there was little evidence of improvement in the outcome for epileptic patients despite the introduction of many major new antiepileptic drugs which are still in widespread use today. However, the studies reviewed by Rodin were all hospital or institution based and the populations examined were unfavourably biased towards large numbers of chronic drug resistant patients.

In the last decade there has been a considerable change in our understanding of the prognosis of epilepsy. My colleagues and I have shown in long-term studies in both adults and children with two or more previously untreated tonic-clonic, or partial ± tonic-clonic seizures that carefully moni-

TABLE 1

PROGNOSTIC STUDIES OF NEWLY DIAGNOSED PATIENTS

	Number	Duration of remission (years)	% in remission
COMMUNITY (retrospective)			
Annegers et al[10]	457	5	70
Goodridge and Shorvon[11]	122	4	69
HOSPITAL (retrospective)			
Okuma and Kumashiro[9]	1868	3	58.3
HOSPITAL (prospective)			
Elwes et al[3]	106	2	82
Beghi and Tognoni[8]	283	From Treatment*	48

*Mean follow up 21.6 months.

tored monotherapy with several of the currently available antiepileptic drugs is associated with at least one or two year remission in approximately 70%–80% of patients[5,6,7]. In a multicentre Italian study nearly one half of 283 newly diagnosed epileptic children and adults (including 51 with single seizures) remained seizure free at a mean of 21.6 months after the start of monotherapy[8]. Similarly a retrospective hospital based study in Japan[9] and retrospective community based or general practitioner based studies in the USA[10] and UK[11] have all revealed a much better prognosis than had been suspected from Rodin's review (Table 1). The key to understanding the much more favourable prognosis revealed in the recent reports is that the latter are all based on the study of newly diagnosed patients followed from the onset of their disorder, in contrast to the previous studies of retrospective groups of patients of varying chronicity. It is now seen that some three-quarters of all patients can enter prolonged remission and in many of these patients antiepileptic drugs can eventually be withdrawn.

The process of epilepsy

The question arises why do some patients with newly diagnosed epilepsy develop chronic epilepsy while the majority do not[1,12,13]. Is the evolution of chronic epilepsy unavoidable because of inherently 'severe' epilepsy which is therefore too powerful for currently available drugs, or is there a process of escalation of epilepsy which can be prevented by early, more effective treatment at the onset of the disorder? Although these two views are not mutually exclusive, my colleagues and I have argued from our own and other evidence that epilepsy should be viewed as a process which can escalate out of control unless 'arrested' early[1,12,13,14,15]. It was Gowers[16] in fact who first hinted at this concept when he proposed that seizures may beget further seizures:

'When one attack has occurred, whether in apparent consequence of an excitant or not, others usually follow without any immediate traceable cause. The effect of a convulsion on the nerve centres is such as to render the occurrence of another more easy, to intensify the predisposition that already exists. Thus every fit may be said to be, in part, the result of those that have preceded it, the cause of those which follow it.'

Gowers presented evidence from his own careful observations that the prognosis for seizure control was inversely proportional to the duration of the disorder. He emphasised the very favourable prognosis in patients with a seizure disorder of less than one year (83% 'arrested') and the relatively high probability that seizures would not be controlled if the disorder had been present for more than 5 years. Gowers' views and their implications were overlooked for most of this century, as the history of drug trials and treatment until quite recently testifies.

Ideally, the question of 'severe' or evolving epilepsy should be studied prospectively in untreated patients but, as already noted, for ethical reasons the natural history of untreated epilepsy is unknown. Elwes et al[3] have attempted a limited retrospective study in patients presenting to the neurology department at King's College Hospital with between two and five untreated tonic-clonic seizures. They examined the time intervals between each seizure in 183 patients in whom it was possible to accurately date the attacks. This was possible in many patients presenting with the more dramatic tonic-clonic seizures but is rarely possible in patients presenting with partial seizures, many of which are uncountable or even unknown to the patient. Overall, they found a steadily declining time interval between attacks, whether patients presented with three, four or five seizures. In only 20% of patients were there examples of lengthening between two successive time intervals. Such a study has several methodol-

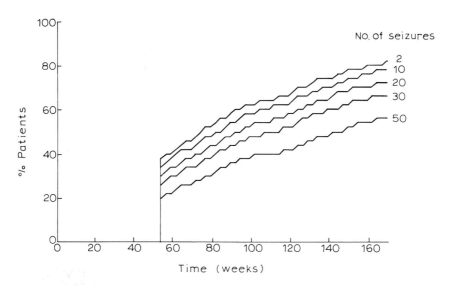

Fig. 1. Influence of pre-treatment seizure number on one year remission rates (Cox's Proportional Hazards Model)[6].

ogical weaknesses related to its retrospective, hospital-based nature, dictated by ethical considerations. Nevertheless, the findings are in keeping with the concept of an escalating process of epilepsy, at least in its early stages, and again emphasise the need to find some way of examining the issues prospectively.

In the last decade my colleagues and I have undertaken two separate prospective studies of the comparative efficacy and toxicity of four major antiepileptic drugs (phenobarbitone, phenytoin, carbamazepine and sodium valproate) in 243 adults and 167 children with newly diagnosed previously untreated epilepsy (at least two tonic-clonic ± partial seizures)[6,7]. In both studies we found no significant differences in efficacy between the four drugs and, as already noted above, the overall prognosis with each of the drugs was very good in both adults and children. By combining the data from each of the four drug groups we were able to examine those factors other than drug therapy which influence long-term prognosis[17]. Remarkably, in both adults and children,

the most significant prognostic factor was the number of seizures prior to treatment; the more untreated seizures the worse the prognosis, as measured either by time to first seizure recurrence after treatment, or number of patients entering one year remission. This is illustrated in Figure 1 which shows the one year remission rates in the adult study. The Figure is a model based on the data, using the Cox's Proportional Hazards Analysis, and shows the declining one year remission curves with increasing number of seizures prior to treatment. By combining the data for the adults and children we were also able to show that the phenomenon can be demonstrated for tonic-clonic or partial seizures alone respectively. Thus it appears that for the best outcome (as far as efficacy but not toxicity is concerned) the timing of the initiation of therapy may be more important than the actual choice of drug. This is perhaps the most powerful evidence so far in support of our view that early treatment may be important to 'arrest' epilepsy and prevent chronic epilepsy.

Relevance of kindling

Is the process of evolving epilepsy which I have discussed related in any way to the phenomenon of kindling[15]? There must be some doubt about this if kindling is defined strictly in terms of the experimental (and possible clinical) effects of repeated subclinical stimuli which culminate in the occurrence of the first seizure. The events I have been concerned with occur after the first seizure. Is this an extension of kindling or a completely different phenomenon? Although recurrent spontaneous seizures have been described in the kindling model in some species this aspect seems to have received less attention. Furthermore, for the most part kindling studies are undertaken in animals which are not genetically predisposed to spontaneous seizures, whereas a genetic background of varying degree is widely suspected in clinical epilepsy.

For the time being it seems appropriate to consider the 'Gowers' Phenomenon', as I have called it[18], as separate and distinct from kindling even though the analogies between the two are readily apparent and probably would have fascinated Gowers. Both take a longitudinal view of seizure processes over time. In the case of kindling it is the impact of suitably spaced successive subclinical stimuli (SSSSS) that is the key. In the Gowers' Phenomenon we are concerned with the effect of suitably spaced successive seizures (SSSS). Exactly how each stimulus leaves its mark on the nervous system in kindling is unknown but if a subclinical stimulus can leave its unidentified but undoubted imprint is it not reasonable to suppose that a seizure itself can do something similar[13,15]? Perhaps both phenomena have more in common with memory and learning. In the former the brain is 'learning' to develop a seizure, in the latter 'learning' to have more seizures. In both processes 'inhibitory' phenomena come into play which is why the spacing of the stimuli or the seizures is important.

Remission in epilepsy

There seems little doubt that some forms of epilepsy, especially in childhood, may go into spontaneous remission. Petit mal and benign Rolandic epilepsy of childhood are classic examples. Perhaps maturational factors in the nervous system are important, but it is also plausible that in childhood and in the adult the brain may generate processes of inhibition as well as the putative excitatory mechanisms I have discussed.

An interesting experimental phenomenon is the resistance to seizures that can be demonstrated immediately following a convulsion[19,20]. Clinically it reminds one of those epileptic subjects who, following a seizure or bout of seizures, can be reasonably confident that they will not have another attack for a certain interval, which varies between patients, but is usually consistent in the individual. A variant which I have observed is the patient with frequent, perhaps daily complex partial seizures who, following a rare tonic-clonic seizure, may not have another complex partial seizure for a month or more. The tonic-clonic seizure seems to clear the system and prevent the minor attacks at least temporarily. Hughlings Jackson[21] was aware of this phenomenon when he commented:

'The other day I congratulated a mother on the fact that her son had not had a severe fit. She however regretted it, saying that the severe fit 'cleared the system', whilst the slight fits rendered him from their frequency unable to go to business.'

An important but unanswered question is whether antiepileptic drug therapy, apart from suppressing seizures during the period of administration, actually enhances the natural process of remission that obviously occurs in some patients. In this regard it is interesting that in a multicentre study of antiepileptic drug withdrawal the duration of remission prior to withdrawal was an important factor favouring successful outcome, i.e. the longer the remission period the more successful the outcome of withdrawal[22].

Conclusions

Epilepsy has been viewed as a symptom, a syndrome (or syndromes) or an illness. An additional or alternative view is that epilepsy is a process, which changes its character over time. We know very little about the natural history of untreated epilepsy, but there is evidence in some patients with tonic-clonic seizures of an accelerating process in the early stages of the disorder, perhaps by the mechanism proposed by Gowers. Recent studies have revealed a much more favourable prognosis for newly diagnosed epileptic patients than had hitherto been reported from the study of chronic patients. Only approximately 20% to 30% go on to develop chronic epilepsy. Longer term prognosis is influenced by the number of seizures prior to treatment. If seizures are not 'arrested' (to use Gowers' terminology) within two years of the onset of the disorder and treatment, there is a considerable risk of chronic drug resistant epilepsy. The contrast between the efficacy of the same drugs in newly diagnosed and in chronic patients is striking. In the latter stages of epilepsy new drugs or alternative strategies of treatment (for example, surgery) are needed. Gowers thought that the spontaneous remission of seizures was unusual, but this clearly occurs in some patients, especially perhaps in certain childhood disorders. The brain is after all the seat of competing excitatory and inhibitory processes and it is possible for the latter to dominate.

This more dynamic longitudinal view of epilepsy has certain implications which require further study: (1) The possibility exists that early effective treatment in the newly diagnosed patient including those with single unprovoked seizures, may prevent ('arrest') the evolution of chronic drug resistant epilepsy. Studies of the effect of early versus delayed treatment strategies on longer term prognosis would shed light on this possibility, as well as on the natural history of untreated epilepsy. (2) Treatment strategies will need to vary according to the stage in the evolution of the disorder that has been reached. (3) The evaluation of new drugs for epilepsy will need to take account of the stage in the evolution of epilepsy on which they are being assessed. (4) The question arises whether antiepileptic drug therapy, in addition to suppressing seizures, will increase the prospects for spontaneous remission.

References

1 Reynolds, E.H. (1987) Epilepsia 28: 97–106.
2 Shorvon, S.D. (1987) In: Warlow, C., Garfield, J. (eds), More Dilemmas in the Management of the Neurological Patient. Churchill Livingstone, Edinburgh, pp 8–13.
3 Elwes, R.D.C., Johnson, A.L., Reynolds, E.H. (1987) Br. Med. J. 297: 948–50.
4 Rodin, E.A. (1968) The prognosis of patients with epilepsy. Charles C. Thomas, Springfield Il.
5 Elwes, R.D.C., Johnson, A.L., Shorvon, S.D., Reynolds, E.H. (1984) N. Engl. J. Med. 311: 944–7.
6 Heller, A.J., Chesterman, P., Elwes, R.D.C., Crawford, P., Chadwick, D., Johnson, A.L., Reynolds, E.H. (1989) Epilepsia 30: 648.
7 De Silva, M., McArdle, B., McGowan, M., Neville, B.G.R., Johnson, A.L., Reynolds, E.H. (1989) Epilepsia 30: 662.
8 Beghi, E., Tognoni, G. (1988) Epilepsia 29: 236–43.
9 Okuma, T., Kumashiro, H. (1981) Epilepsia 22: 35–53.
10 Annegers, J.F., Hauser, W.A., Elveback, L.R. (1979) Epilepsia 20: 729–37.
11 Goodridge, D.M.G., Shorvon, S.D. (1983) Br. Med. J. 287: 641–7.
12 Reynolds, E.H., Elwes, R.D.C., Shorvon, S.D. (1983) Lancet 2: 952–4.
13 Reynolds, E.H. (1988) Epilepsia 29(Suppl.1): 525–8.
14 Shorvon, S.D., Reynolds, E.H. (1986) In: Trimble, M.R., Reynolds, E.H. (eds.) What is Epilepsy? Edinburgh, Churchill Livingstone, pp 36–45.
15 Reynolds, E.H. (1989) In: Bolwig, T.G., Trimble, M.R., (eds.) The Clinical Relevance of Kindling. Chichester. John Wiley, pp 149–60.

16 Gowers, W.R. (1881) Epilepsy and Other Chronic Convulsive Diseases. London: Churchill.

17 Reynolds, E.H., Heller, A.J., Elwes, R.D.C., de Silva, M., Neville, B., Chadwick, D., Johnson, A.L. (1989) Epilepsia 30(5): 648.

18 Reynolds, E.H. (1981) In: Reynolds, E.H., Trimble, M.R. (eds) Epilepsy and Psychiatry. Churchill Livingstone, Edinburgh, pp 264–290.

19 Herberg, L.J., Tress, K.H., Blundell, J.E. (1969) Brain 92: 313–328.

20 Green, A.R., Nutt, D.J., Cowan, P.H. (1981). In: Sandler, M. (ed) The Psychopharmacology of Anticonvulsants. Oxford University Press, Oxford.

21 Jackson, J.H. (1873) On the anatomical, physiological, and pathological investigation of epilepsies. Rep. West Riding Lunatic Asylum, 3: 315–339.

22 Chadwick, D. Personal communication.

Experimental Aspects

New Antiepileptic Drugs (Epilepsy Res. Suppl. 3)
Editors: F. Pisani, E. Perucca, G. Avanzini, A. Richens
© 1991 Elsevier Science Publishers B.V. (Biomedical Division)

CHAPTER 3

Rational basis for the development of new antiepileptic drugs

Roberto Mutani, Roberto Cantello, Maria Gianelli, Diego Bettucci

Department of Neurology, Turin University School of Medicine at Novara, Ospedale Maggiore, Novara, Italy

Introduction

For the 50 million people of the world suffering from epilepsy there are a large number of drugs on the market. Unfortunately, 20–30% of patients are refractory to treatment with the currently available antiepileptic armamentarium, even if diagnosis and treatment are optimally managed[1]. New antiepileptic drugs should be strongly investigated in preclinical studies and the most promising ones should proceed to clinical trials. At variance with the often serendipitous development of conventional anticonvulsants, the rational approach to the development of new drugs should be based on knowledge of the basic events responsible for epilepsy[2].

There is a large body of evidence[3] that the electrophysiological pattern of focal epileptogenesis (the paroxysmal depolarization shift with burst firing) can be generated under three main conditions: i) impairment of GABA-mediated inhibition; ii) enhancement of glutamate/aspartate (Glu/Asp) transmission; iii) endogenous bursting characteristics of subsets of pyramidal neurons in the hippocampus and neocortex.

Hypotheses to explain generalized epilepsy are less substantiated. Studies on feline generalized epilepsy[4,5] suggest that the underlying mechanisms might be different from those of focal epilepsy. The ideal condition for the generation of spike-and-wave (SW) pattern would be an enhancement of both Glu/Asp-mediated excitation and GABA-mediated inhibition. Suppression of GABAergic inhibition would promote the transition from non-convulsive generalized attacks to convulsive generalized attacks.

Treatment strategies should be based on analyses of the reported cellular mechanisms. Pharmacological manipulation of these mechanisms provides the rational basis for the development of new drugs[6].

Impairment of GABA-mediated transmission and its pharmacological manipulation

GABA, synthetized in the brain from glutamic acid by decarboxylation (GAD) and metabolized by transamination (GABA-T), is the major inhibitory neurotransmitter. It acts via $GABA_A$ receptors to open Cl^- channels and hyperpolarize the membrane, and via $GABA_B$ receptors to open K^+ or Ca^{++} channels[7]. Impairment in the function of GABA neurons leads to seizures: compounds which diminish GABA synthesis (isoniazid, allylglycine) or which impair GABA receptor function (bicuculline, penicillin, picrotoxin, some beta-carbolines) induce convulsions[8]. Conversely, agents that decrease the metabolic breakdown of GABA (valproate) or which facilitate GABA-mediated events by interaction with the GABA receptor complex (barbiturates, benzodiazepines) are anticonvulsant agents.

The development of cobalt and alumina foci in neocortex is paralleled by a decrease in all aspects of GABAergic transmission (GAD activity, GABA levels, GABA uptake)[9,10,11,12]. More recently, Ribak et al.[13] provided an elegant demonstration that a selective decrease in the number of GABAergic cells occurs in pre-seizing monkeys implanted with alumina gel. Though not confirmed by other investigators[14], neurochemical findings indicate an alteration of GABAergic neurons in epileptic tissue specimens of refractory temporal lobe epileptic patients[15].

GABA-mediated inhibition can be enhanced by a variety of mechanisms. GABA itself does not cross the blood/brain barrier. The technique of incorporating GABA into liposomes that cross the blood/brain barrier has been used with success in some models[16] but is not yet available for clinical use. True GABA agonists acting at $GABA_A$ (muscimol, THIP) and $GABA_B$ receptors (baclofen) show anticonvulsant action in some rodent systems, but in clinical trials they lack convincing therapeutic activity and show toxicity[8]. In primates these compounds produce petit-mal like seizures and enhance epileptic activity in models of absence seizures, confirming the hypothesis that an active GABAergic tone is the prerequisite for generating SW discharges[4].

GABA prodrugs cross the blood/brain barrier and are metabolized to GABA in the brain. Progabide shows a wide spectrum of action in models related (bicuculline, penicillin, picrotoxin, allylglycine) and unrelated (kainate, strychnine, Papio-Papio, ESh) to impairment of GABAergic transmission[17].

The anticonvulsant beta-carbolines act at the GABA receptor complex and produce, through allosteric enhancement, the potentiation of the postsynaptic action of GABA by increasing the number of Cl^- channels opened by a given concentration of GABA. Beta-carbolines show anticonvulsant activity in some rodent and primate models of epilepsy when given acutely, and lack sedative and muscle relaxant effects[6].

Stimulation of GABA synthesis can be obtained by enhancement of GAD activity. Milacemide elevates GAD activity, mainly in substantia nigra (30%), and shows antiepileptic action in various models, including the mongolian gerbil (in this model epileptic activity has been correlated to decreased GABAergic activity in substantia nigra)[18].

An increase in brain GABA levels can be obtained by inhibition of GABA-T. Vigabatrin is a potent, selective and irreversible inhibitor of GABA-T. It induces a dose-dependent increase in brain GABA levels and shows anticonvulsant action in models related and unrelated to GABAergic impairment[19]. Injected into the substantia nigra and area tempestas [20], vigabatrin suppresses the limbic and motor seizures in the kindling model, indicating potential efficacy against partial and generalized convulsive seizures.

Inhibition of both neuronal and glial GABA uptake prolongs GABA action. In preclinical studies nipecotic acid, as its ethyl ester or with a lipophilic side-chain, penetrates the blood/brain barrier and shows anticonvulsant action in various rodent models[21]. Stiripentol inhibits both GABA-uptake and GABA-T and also shows anticonvulsant effects in various models[22].

If it is true that impairment of GABA-mediated inhibition is crucial for focal epileptogenesis and seizure generalization, then one would expect that the increase of brain GABA levels, obtained through any of these pharmacological manipulations, is beneficial. However, as pointed out by Gale[23], increased GABA transmission does not always result in inhibition of brain excitability. In some areas of the brain (e.g. caudate nucleus, superior colliculus), the elevation of GABA transmission is actually proconvulsant.

The fact that GABA-elevating drugs exert a net anticonvulsant effect (and this is fortunate for therapeutic purposes) suggests that actions at anticonvulsant sites outweigh the actions at proconvulsant sites. Two brain areas, substantia nigra and area tempestas, are examples of specific brain regions where elevations of GABA can achieve an anticonvulsant effect. Substantia nigra output exerts a proconvulsant effect, lowering convulsive threshold and facilitating discharge propagation, and is inhibited by GABA[24]. The area tempestas participates in the triggering and initiation of seizures associated with limbic forebrain circuits. GABA elevation in the area tempestas prevents seizures in various models[25].

Glu/Asp excitation and its pharmacological manipulation

Systemic or topical application of Glu/Asp produces convulsions in animals[26]. Kainate, an excitotoxic analogue of Glu, induces prolonged status epilepticus[27]. Craig and Hartman[28] found markedly decreased levels of Glu/Asp in the brain of cobalt-treated rats. In the feline generalized penicillin epilepsy model and in the preepileptic period, Van Gelder et al.[29] found a decreased brain content of Glu/Asp, which was interpreted as due to loss from the interneuronal compartment into the extracellular space. Glu has been found to be elevated in the serum of patients with absence seizures[30]. Increased levels of quinolate, an excitotoxic agent proposed as an endogenous N-methyl-D-aspartate (NMDA) receptor agonist, have been reported in epileptic human brain tissue[31].

The receptors for excitatory aminoacids are currently divided into at least three categories, NMDA, kainate and quisqualate[32]. Glu activates all three types of receptor, whereas Asp activates NMDA receptors selectively. NMDA receptors under baseline conditions are blocked by physiological concentrations of Mg^{++}. This block is voltage-dependent and, when neurons are depolarized, they become responsive to NMDA agonists. Thus it has been postulated that NMDA receptors may be more involved in epileptiform activity than in background synaptic activity[3]. Also, NMDA receptor activation plays a crucial role in long-term potentiation[33], learning and memory[34] and kindling[35]. An excessive NMDA receptor activation with massive Ca^{++} influx is thought to occur during repeated seizures and brain ischemia[36], and would be responsible for neuronal damage and the long-lasting plastic changes reinforcing the epileptogenic condition as seizures keep recurring[5].

Impairment of excitatory aminoacid transmission can be achieved by inhibition of synaptic release. Lamotrigine is a potent inhibitor of release, with a phenytoin-like pharmacological profile in many model systems[37].

Many NMDA receptor antagonists are presently under preclinical evaluation. Both competitive (e.g. APH) and non-competitive (e.g. MK-801) blockers of NMDA receptor-activated ion channels are anticonvulsant in a wide range of seizure models (amygdala kindling, mice and rats with sound-induced seizures, baboons with photically-induced myoclonus). An important action of NMDA antagonists is cerebroprotection, including prophylaxis and therapy for ischaemic brain damage and for damage following cerebral trauma[38-40].

Modulation of Ca^{++} currents

In vitro studies have shown that the paroxysmal depolarization shift with burst firing is an intrinsic property of some pyramidal neurons in the CA_3 area of hippocampus and layer 4 of neocortex[41]. This property would explain on the one hand the high epileptogenic susceptibility of these brain regions and on the other hand, as the paroxysmal depolarization shift is mainly due to Ca^{++} influx, their vulnerability to repeated seizures[36]. The hypothesis has been put forward that neurons in these areas have either a higher concentration of voltage-sensitive Ca^{++} channels or a lower concentration of opposing K^+ channels[3].

In preclinical trials the Ca^{++} antagonists flunarizine, verapamil and nimodipine have shown anticonvulsant activities in various models of epilepsy[42–44]. Different Ca^{++} currents exist, each of which may be involved in epilepsy. Thus, it would be desirable to develop drugs capable of crossing the blood brain barrier and selective for each of the following Ca^{++} currents: the voltage-sensitive, the synaptic terminal and the neurotransmitter (NMDA)-activated current. It would also be useful to have drugs capable of enhancing extrusion of Ca^{++} from neurons, thus avoiding Ca^{++} accumulation intracellularly and its adverse consequences[3].

Antiepileptogenic and antiepileptic properties of the new antiepileptic drugs

Epileptologists have long recognized that preventing epilepsy would be preferable to chronic therapy aimed at controlling seizures after epilepsy has developed. In this perspective, the strategies for developing new drugs on the basis of their potential capability of suppressing only epi-

leptic activity appear to be limited. The ideal new drugs should also be able to block the events, unfortunately mostly unknown, that from a potentially epileptogenic brain lesion lead to the development of epilepsy. Available drugs do not necessarily show both antiepileptogenic and antiepileptic actions. An example of this discrepancy is shown in the kindling model[45–47] (Table 1).

The data shown in Table 1 indicate that some drugs preferentially show antiepileptogenic or antiepileptic action and that only some show both actions. Despite our incomplete understanding of the basic mechanisms of kindling[48–50] and the unproven occurrence of kindling in man, this model is recommended to test the antiepileptogenic and antiepileptic properties of new drugs.

Endogenous anticonvulsants

Recent research has shown that some endogenous anticonvulsant substances are released in the brain during seizures, as a possible neurochemical adaptive process that terminates seizures and prevents further attacks (e.g. opioid peptides, cholecystokinin, adenosine and purines, the quinolinic-antagonist kynurenic acid)[2,51–53].

A rewarding strategy for future research could be the drug-induced potentiation of these endogenous mechanisms. Also, if some forms of epilepsy were found to be due to a defect of a specific

TABLE 1

THE ACTION OF OLD AND NEW ANTIEPILEPTIC DRUGS ON DEVELOPMENT OF KINDLING AND ON KINDLED SEIZURES

Drug	Development of kindling	Kindled seizures
Phenobarbitone, valproic acid, benzodiazepines	+	+
Phenytoin, carbamazepine	−	+
GABA-agonists	+	+
Lamotrigine	+	+
MK-801	+	−
Alpha$_2$-agonists	+	−

endogenous anticonvulsant system, a new form of treatment, i.e. replacement therapy, could be considered as optimal.

Concluding remarks

Though rationally developed on the basis of present knowledge of the cellular mechanisms of epilepsy, the recently tested new drugs do not seem to have achieved the expected optimal results in patients resistant to conventional anticonvulsants[54]. This could be due to the following reasons:

1. Though empirically developed, conventional drugs already act on some of the cellular mechanisms of epileptogenesis[2,55,56]. Phenobarbitone, valproic acid and benzodiazepines enhance GABA-mediated inhibition. Phenytoin, carbamazepine and valproic acid decrease Na^+ conductance, thus impairing repetitive neuronal firing. Excitatory aminoacid transmission is decreased by phenobarbitone, phenytoin, carbamazepine, valproic acid and benzodiazepines. Phenobarbitone, phenytoin and carbamazepine block voltage-dependent Ca^{++} entry into neurons. Phenytoin, carbamazepine and benzodiazepines act on adenosine receptors and/or uptake systems. Thus, there would be no real qualitative difference between the conventional and already available new drugs. New rational approaches are needed for developing the future new antiepileptic drugs.

2. The already-available new drugs have been mostly tested in severe refractory patients with a long history of partial and/or generalized convulsive seizures. This might explain their lower than expected degree of efficacy. If it is true that repetition of seizures causes permanent plastic effects with progressive reinforcement of both the epileptic condition and drug resistance[5], then a real evaluation of the antiepileptic efficacy of the new drugs should be better done in patients not yet treated or at an early phase of their history of epilepsy.

References

1 Juul-Jensen, P. (1986) In: Schmidt, D., Morselli, P.L. (eds) Intractable Epilepsy, Experimental and Clinical Aspects. Raven Press, New York, pp 5–11.
2 Mutani, R., Cantello, R., Gianelli, M., Verzè, L. (1988) In: Faienza, C., Prati, G.L. (eds) Diagnostic and Therapeutic Problems in Pediatric Epileptology. Elsevier, Amsterdam, pp 169–177.
3 Dichter, M.A. (1989) Epilepsia 30 (Suppl. 1): S3–S12.
4 Gloor, P., Fariello, R.G. (1988) TINS 11: 63–68.
5 Gloor, P. (1989) Can. J. Neurol. Sci. 16: 8–21.
6 Meldrum, B.S. (1986) In: Meldrum, B.S., Porter, R.J. (eds) New Anticonvulsant Drugs. Libbey, London, pp 17–30.
7 Bormann, J. (1988) TINS 11: 112–117.
8 Fariello, R.G. (1985) In: Porter, R.J., Morselli, P.L. (eds) The Epilepsies. Butterworths, London, pp 1–19.
9 Ribak, C.E., Harris, C.B., Vaughn, J.E., Roberts, E. (1979) Science 205: 211–214.
10 Ross, S.M., Craig, C.R. (1981a) J. Neurochem. 36: 1006–1011.
11 Ross, S.M., Craig, C.R. (1981b) J. Neurosci. 1: 1388–1396.
12 Bakay, R.A., Harris, C.B. (1981) Brain Res. 206: 387–408.
13 Ribak, C.E., Joubran, C., Kesslak, J.P., Bakay, R.A. (1989) Epilepsy Res. 4: 126–138.
14 Sherwin, A.L., Van Gelder, N.M. (1986) In: Delgado-Escueta, A.V., Ward, A.A., Woodbury, D.M., Porter, R.J. (eds) Basic Mechanisms of the Epilepsies. Molecular and Cellular Approaches. Raven Press, New York, pp 1011–1032.
15 Lloyd, K.G., Bossi, L., Morselli, P.L., Munari, C., Rougier, M., Loiseau, H. (1986) In: Delgado-Escueta, A.V., Ward, A.A., Woodbury, D.M., Porter, R.J. (eds) Basic Mechanisms of the Epilepsies. Molecular and Cellular Approaches. Raven Press, New York, pp 1033–1044.
16 Loeb, C., Besio, G., Mainardi, P., Scotto, P., Benassi, E., Bo, G. (1986) Epilepsia 27: 98–102.
17 Morselli, P.L., Bartholini, G., Lloyd, K.G. (1986) In: Meldrum, B.S., Porter, R.J. (eds) New Anticonvulsant Drugs. Libbey, London, pp 237–254.
18 Porter, R.J. (1989) Epilepsia 30 (Suppl. 1): S29–S34.
19 Schechter, P.J. (1986) In: Meldrum, B.S., Porter, R.J.

(eds) New Anticonvulsant Drugs. Libbey, London, pp 265–276.

20 Gale, K. (1988) Epilepsia 29 (Suppl. 2) : S15–S34.

21 Yunger, L.M., Fowler, P.J., Zarevics, P., Setler, P.E. (1984) J. Pharmacol. Exp. Ther. 228: 109–115.

22 Vincent, J.C. (1986) In: Meldrum, B.S., Porter, R.J. (eds) New Anticonvulsant Drugs. Libbey, London, pp 225–264.

23 Gale, K. (1989) Epilepsia 30 (Suppl. 3): S1–S11.

24 Garant, D.S., Gale, K. (1987) Exp Neurol 97: 143–159.

25 Piredda, S., Pavlick, M., Gale, K. (1987) Epilepsy Res. 1: 102–106.

26 Johnston, G.A. (1972) Biochem. Pharmacol. 22: 137–149.

27 Ben Ari, Y., Trembley, E., Lagowska, J., Le Gal Lassalle, G. (1979) Brain Res. 163: 176–179.

28 Craig, C.R., Hartman, E.R. (1973) Epilepsia 14: 409–414.

29 Van Gelder, N.M., Siatitas, I., Menini, C., Gloor, P. (1983) Epilepsia 24: 200–213.

30 Van Gelder, N.M., Janjua, N.A., Metrakos, K., MacGibbon, B., Metrakos, J.D. (1980) Neurochem. Res. 5: 659–671.

31 Feldblum, S., Rougier, A., Loiseau, H., Loiseau, P., Cohadon, F., Morselli, P.L., Lloyd, K.G. (1988) Epilepsia 29: 23–29.

32 Watkins, J.C. (1989) In: Watkins, J.C., Collingridge, G.L. (eds) The NMDA Receptor. Irl Press, Oxford, pp 1–18.

33 Collingridge, G.L., Davies, S.N. (1989) In: Watkins, J.C., Collingridge, G.L. (eds) The NMDA Receptor. Irl Press, Oxford, pp 123–136.

34 Morris, R.G.M., Davis, S., Butcher, S.P. (1989) In: Watkins, J.C., Collingridge, G.L. (eds) The NMDA Receptor. Irl Press, Oxford, pp 137–152.

35 Cain, D.P. (1989) TINS 12: 6–9.

36 Meldrum, B.S. (1988) Curr. Opin. Neurol. Neurosurg. 1: 563–568.

37 Binnie, C.D., Debets, R.M.C., Engelsman, M., Meijer, J.W.A., Meinardi, H., Overweg, J., Peck, A.W., Van Wieringen, A., Yuen, W.C. (1989) Epilepsy Res. 4: 222–229.

38 Meldrum, B.S., Chapman, A.G., Patel, S., Swan, J. (1989) In: Watkins, J.C., Collingridge, G.L. (eds) The NMDA Receptor. Irl Press, Oxford, pp 207–216.

39 Iversen, L.L., Woodruff, G.L., Kemp, J.A., Foster, A.C., McKernan, R., Gill, R., Wong, E.H.F. (1989) In: Watkins, J.C., Collingridge, G.L. (eds) The NMDA Receptor. Irl Press, Oxford, pp 217–226.

40 Albers, G.W., Goldberg, M.P., Choi, D.W. (1989) Ann. Neurol. 25: 398–403.

41 Prince, D.A., Connors, B.W. (1986) In: Delgado-Escueta, A.V., Ward, A.A., Woodbury, D.M., Porter, R.J. (eds) Basic Mechanisms of the Epilepsies. Molecular and Cellular Approaches. Raven Press, New York, pp 38–49.

42 Wauquier, A., Ashton, D. (1985) In: Speckmann, E., Walden, S., Witte, P. (eds) Epilepsy and Calcium. Urban and Schwarzenberg, Munich, pp 38–49.

43 Walden, J., Speckman, E., Witte, P. (1985) EEG Clin. Neurophysiol. 61: 299–309.

44 Morocutti, C., Pierelli, F., Sanarelli, E., Stefano, A., Peppe, A., Mattioli, G.L. (1986) Epilepsia 27: 498–503.

45 McNamara, J.O. (1989) Epilepsia 30 (Suppl. 1): S13–S18.

46 Williamson, J.M., Lothmann, E.W. (1989) Ann. Neurol. 26: 85–90.

47 Schmutz, M., Klebs, K. (1989) In: Bolwig, T.G., Trimble, M.R. (eds) The Clinical Relevance of Kindling, John Wiley, London, pp 35–54.

48 McNamara, J.O. (1988) TINS 11: 33–36.

49 Racine, R.J., Ivy, G.O., Milgram, N.W. (1989) In: Bolwig, T.G., Trimble, M.R. (eds) The Clinical Relevance of Kindling. John Wiley, London, pp 15–34.

50 Wasterlain, C.G., Morin, A.M., Denson, G., Bronstein, J.M. (1989) In: Bolwig, T.G., Trimble, M.R. (eds) The Clinical Relevance of Kindling. John Wiley, London, pp 55–68.

51 Delgado-Escueta, A.V., Ward, A.A., Woodbury, D.M., Porter, R.J. (1986) In: Delgado-Escueta, A.V., Ward, A.A., Woodbury, D.M., Porter, R.J. (eds) Basic Mechanisms of the Epilepsies. Molecular and Cellular Approaches. Raven Press, New York, pp 3–55.

52 Lapin, I.P., Prakhie, I.B., Kiseleva, I.P. (1986) J. Neural. Transm. 65: 177–185.

53 Chin, J.H. (1989) Ann. Neurol. 26: 695–698.

54 Porter, R.J. (1989) Epilepsia 30 (Suppl. 1): S29–S34.

55 DeLorenzo, R.J. (1988) Epilepsia 29 (Suppl. 2): S35–S47.

56 MacDonald, R.L. (1989) Epilepsia 30 (Suppl. 1) : S19–S28.

New Antiepileptic Drugs (Epilepsy Res. Suppl. 3)
Editors: F. Pisani, E. Perucca, G. Avanzini, A. Richens
© 1991 Elsevier Science Publishers B.V. (Biomedical Division)

CHAPTER 4

Genetic animal models for generalized non convulsive epilepsies and new antiepileptic drugs

Giuliano Avanzini[1], Christian Marescaux[2]

[1]*Istituto Neurologico C. Besta, Milano, Italy;* [2]*Clinique Neurologique C.H.U., Strasbourg, France*

Epileptic syndromes characterized by non convulsive seizures include different forms of infantile epilepsies which are listed in the ILAE classification of 1989[1] as: benign myoclonic epilepsy in infancy, childhood and juvenile absence epilepsies, West syndrome, juvenile myoclonic epilepsy, Lennox-Gastaut syndrome, epilepsy with myoclonic-astatic seizures and epilepsy with myoclonic absences. Convulsive manifestations may also occur in the course of the above mentioned epilepsies, but, with the notable exception of tonic seizures in Lennox Gastaut syndrome, they are not included among the classification criteria. Excluded from this list are early myoclonic encephalopathy and Otahara syndrome (early infantile epileptic encephalopathy with suppression burst) which present with polymorphic seizures and may be the expression of progressive neurological diseases, besides other forms included under the heading 'epilepsies undetermined as to whether they are focal or generalized'. On the contrary the category of 'epilepsies with seizures precipitated by specific modes of activation' is to be taken into consideration with regard to some epileptic syndromes characterized by generalized, mainly non convulsive seizures (i.e. reflex epilepsies with myoclonic or myoclonoastatic seizures provoked by somatosensory stimulation, startle epilepsy with short myoclonus-like seizure, photogenic epilepsies with absences with or without myoclonic component provoked by light stimulation or eye closure).

Overall the different epilepsies with generalized non convulsive seizures include a high proportion of idiopathic, presumably genetic forms, hence the interest in finding animal strains presenting with genetically determined non convulsive seizures to be used as natural models of generalized non convulsive human epilepsies for genetic, pathophysiological and pharmacological investigations. According to Löscher [2,3] an ideal model for epilepsy should fulfill the following requirements: a) liability to spontaneously recurrent seizures similar to those occurring in human epilepsies associated with epileptic-like EEG activity; b) pharmacokinetics of antiepileptic drugs, similar to those in man; c) effective plasma concentrations of antiepileptic drugs similar to those required for control of human epilepsies. Such a model, however, would meet the main criteria upon which seizures[4], but not the epilepsies classification[1] are based.

It must be remembered that according to the ILAE classification the electroclinical characteristics of the seizures are only a part of the cluster of signs and symptoms which allow the definition of an epileptic syndrome. These include such items as 'etiology, anatomy, precipitating factors, age of onset, severity, chronicity, diurnal and circadian cycling and sometimes prognosis'[1]. If all these criteria are to be taken into account, the identification of suitable animal models for human epilepsies becomes obviously more and more problematic.

Genetic animal models for non convulsive generalized epilepsies

Detailed reports of animal strains presenting with spontaneous epileptic seizures have been published in a number of papers. Comprehensive reviews may be found in the symposia reports published on Fed Proc[5,6] and in Löscher[2] and Löscher and Schmidt[3]. The present description will be limited to some rodent and primate strains presenting with non convulsive generalized epilepsies.

Single locus mutants mice. The tottering mouse

This mutant originally described by Green and Sidman[7] is a recessive homozygous with respect to locus tg (Chromosome 8). A neurological deficit consisting in broad based, ataxic gait is first apparent at 3 to 4 weeks of age. As early as 32 days postnatally bilaterally, synchronous 6–7 Hz spike-wave (SW) discharges appear as spontaneous 0.3–10 sec. bursts in electrocorticographic recordings[8]. The SW bursts occur hundreds of times per day and are systematically accompanied by behavioural arrest with vibrissal twitching and single myoclonic jerks of the head or jaw[9]. Their incidence is increased immediately prior to sleep and decreased during alerting stimulations. In addition, spontaneous partial motor seizures consisting of jerks of a single limb with preserved behavioural responsiveness and no EEG correlates are observed by 3–4 weeks of age. A selective overgrowth of locus coeruleus axons raising noradrenaline levels within most terminal fields has been demonstrated by Levitt

and Noebels[10]. Selective neonatal denervation of the hypertrophic noradrenaline arborization by injection of the neurotoxin 6-hydroxydopamine entirely prevents the later appearance of SW seizures in adult animals[11].

Among other single locus mutant mice, the quacking mouse (recessive homozygous, locus qk, chromosome 17) has been reported to exhibit spontaneous and/or stimulus induced myoclonic and generalized tonic-clonic seizures which seem to be related to an hypomyelination disorder[12]. The seizures start with facial twitches and gradually involve the flexor muscles of the neck, trunk and forelimb. They may eventually proceed to a brief, generalized clonic seizure. Myoclonic seizures can be elicited by stress or arousal and are accompanied by electroencephalographic abnormalities[13].

Rats with spontaneous generalized non convulsive epilepsy

Approximately 16–30% of Sprague-Dawley and Wistar rats show spontaneous bilateral and synchronous SW discharges. A strain of Wistar rats with a 100% incidence of SW has been obtained by inbreeding in the Strasbourg Centre de Neurochemie du CNRS[14]. Six to 8 Hz discharges are consistently associated with behavioural manifestation consisting of arrest in movement, twitching of the vibrissae, and single myoclonic jerks of face and neck muscles. Such absence-like seizures occur at a rate of 1 per minute in a quiet awake state and are suppressed during active arousal state. There is no evidence of intercritic EEG discharges (i.e. SW not accompanied by behavioural seizures). SW discharges appear at 40 to 120 days of age and persist all along the animal life span; their duration ranges between 1 and 90 sec (usually about 10 sec with a tendency to increase with age). Rats with non convulsive generalized epilepsy from the Strasbourg colony do not show neurological defects. Neuropathological studies failed so far to show any significant alterations. The mode of genetic transmission is still under evaluation. Recently another Wistar strain with high incidence of spontaneous absence-like seizures has been reported[15].

Mongolian gerbil (Meriones unquiculatus)

This rodent which belongs to the family of Muriadae has been reported by Thiessen et al[16] to exhibit seizures precipitated by environmental changes ('xenogenic')[17], with a particularly high incidence in selected inbred strains[18]. According to Majakowski and Donadio[19] xenogenic seizures, occurring when the animal is put in an open field, start with a prodromic phase characterized by searching behaviour associated with sporadic spikes and sharp waves which gradually increase in frequency, become more synchronized and eventually generalize progressing to a myoclonic phase. This latter is characterized by cessation of ongoing motor activity, crouching, twitches of the ear and eyelids, and isolated body jerks associated with bursts of high voltage, high frequency spikes, followed by EEG depression. The myoclonic phase lasts from 2 to 18 sec and may end abruptly or develop in clonic-tonic (phase 2), tonic (phase 3) and clonic (phase 4) manifestations followed by a post seizure coma[19]. The mean age of onset of seizures was 57 ± 3 days in males, and 47 ± 3 days in females in the study by Loskota et al.[18]. Vascular anomalies in the territory of anterior cerebral arteries of mongolian gerbils have been reported by Berry et al.[20], Schonfeld and Glick[21] and Donadio et al.[22], but their correlation with seizures susceptibility is doubtful[23].

Baboons with photomyoclonic seizures

A photomyoclonic response in the baboon Papio papio from Casamance, Senegal, was first reported by Killam et al. in 1966[24]. Further studies demonstrated that animals from Casamance are more often light sensitive (52.4%) than those from other Senegal areas[25]. In photosensitive animals, 20–30 Hz intermittent light stimulation evokes rhythmic myoclonic movements of the eyelids and periocular musculature, followed by spread of diffuse clonic twitching to the face and neck, often accompanied by more intense, isolated jerks of the head. Finally the entire body may become involved in violent jerks with flexion of the head and upper body and extension of the lower limbs. In a minority of animals (about 20%) the seizure may proceed to a full tonic-clonic seizure. The photomyoclonic seizures are associated with bilateral poly-spike and waves (PSW) in the frontal and fronto-central regions, synchronous with the myoclonic jerks. Jerks and PSW may occur spontaneously in some animals even in the absence of photic stimulation[26]. Spontaneous PSW paroxysms appear through most stages of sleep with a marked increase in REM sleep[26,27]. The photosensitivity becomes manifest after the 5th month of age and lasts through all the life span[25,28] with marked variability over the time. A slight predominance in photosensitivity has been reported in young females with respect to young males[25]. Neuropathologic studies failed to show any changes in the central nervous system of Papio papio. A photomyoclonic response has been reported to occur in a limited percentage of different primate species: Papio cynocephalus from Ethiopia[29], Erythrocebus patas, and Cercopithecus aethiops sabaeus[24].

Relevance of animal models to human non convulsive epilepsies

Attempts to correlate the characteristics of animal epilepsies with the human counterparts have led to propose some specific correspondences. Along this line tottering mice and rats with SW discharges have been proposed as suitable models for human petit mal epilepsy[3,14,30]. Indeed the behavioural arrests with vibrissal twitching associated with SW occurring in these animals are highly reminiscent of human absence seizures. Children with petit mal epilepsy, however, are neurologically normal by definition and the same is true for Wistar rats with generalized SW but not for tottering mice. Both animal models and human petit mal share the important character of the age dependency. However, according to the biological criteria of age correspondence, the age of onset of absence-like seizures in Wistar rats and tottering mice is more mature than in humans. Moreover, in animals the epileptic symptomatology tends to persist unchanged for all life, while human absences may disappear or evolve into a combination of absence and grand mal seizures after puberty. In both Wistar rats and tottering mice the SW frequency is consistently higher than in human petit mal. Finally, in tottering mice absence-like seizures are associated with focal seizures which are never observed in typical human petit-mal. It may be interesting to note however, that in another form of human epilepsy (i.e. benign partial epilepsy with centro temporal spikes) bilateral SW discharges independent from typical focal EEG pattern may occasionally be observed.

Rather unwarranted is the proposed assimilation of the quacking mutant mouse with the human forms of West and Lennox-Gastaut syndromes, based on the presumed common feature of myelin abnormalities as proposed by Seyfried and Glaser[12].

Unlike human epilepsies, animal genetic epilepsies are often characterized by reflex seizures induced by acoustic (audiosusceptible mice and rats), photic (photosensitive fowl and baboon) or complex stimulation (mongolian gerbil). Of the two animal models of genetic epilepsies with reflex seizures included in the present review, the photosensitive Papio papio has been particularly investigated in comparison with human photo-

genic epilepsies.

About 4% of patients with epilepsy are liable to visually-induced seizures, and an even smaller percentage (variable in different studies) can be diagnosed as having photogenic reflex epilepsy, defined as the occurrence of all or many seizures in close relation with flickering light encountered in every day life. The EEG response to rhythmic photic stimulation (photoparoxysmal response) consists, like in the photosensitive baboon, of generalized SW and PSW but, unlike in Papio papio, is best evoked at flicker frequencies between 15 and 20 Hz. Other differences are to be found in seizure characteristics: only few cases of human photogenic epilepsies present with myoclonic seizures or with absences with or without a myoclonic component, while the most frequently observed type of seizure is the generalized convulsive seizure. Based on these narrow criteria, the photosensitive Papio papio can therefore be considered a suitable model for idiopathic generalized epilepsies occurring only exceptionally in man.

Even more problematic is the assimilation of human and gerbil epilepsies. Although seizures precipitated by complex environmental stimuli are anecdotally reported in man, they can hardly be equated to the xenogenic seizures in gerbils.

It can be concluded that none of the animal models considered here is strictly superimposable to a human form. This conclusion does not rule out the possibility that species-related differences may lead to a different expression of common physiopathological mechanisms, which can be profitably investigated in animals.

Information from animal models on pathophysiology of generalized non convulsive epilepsies

The classic question of the cortical versus subcortical origin of the discharge underlying generalized seizures has been addressed to the Papio papio model as early as the sixties[26,31]. Since then much work has been devoted to the neurophysiology of this model (see a recent review by Menini and Silva Barrat[32]), leading to the following conclusions. The photosensitivity of the Papio papio does not result from a dysfunction of the specific visual system but from a combination of a photically-induced facilitation of peristriatal neurons with a hyperexcitable state of the fronto-rolandic area, which receives cortico-cortical afferences from the occipital cortex. The involvement of subcortical structures (i.e. thalamic nuclei) in the photogenic epileptic discharge is only secondary to the cortical activation[31,33]. A primary role of an hyperexcitable state of the neocortex for the generation of epileptic discharges is also supported by results obtained in mongolian gerbils[19], while in tottering mice and Wistar rats with non convulsive generalized epilepsy the role of the neocortex as promoter of SW discharge is still to be defined. As reported above, the cellular expression of tg mutation in mice is a hypertrophy of the locus coeruleus axonal plexus resulting in noradrenergic hyperinnervation of terminal fields[10]. Since the locus coeruleus projections involve both subcortical (i.e. thalamic) and cortical areas, its supposed epileptogenic effect on target neurons may be exerted at either level. Moreover, in Wistar rats with non convulsive generalized epilepsy, thalamic nuclei have been shown to be greatly involved in SW discharges since their onset. Experiments carried out in our laboratories in the context of a mutual collaborative study, have provided preliminary evidence of a critical role of the Nucleus Reticularis Thalami (RTN) in regulating SW discharge in Wistar rats. RTN neurons are characterized by peculiar Ca^{2+} and Ca^{2+}- dependent K^+ conductances which enable them to produce 6–8 Hz oscillatory activities[34]. Because of these intrinsic properties and of its anatomical connectivities (it receives cortico-

thalamic projections and is reciprocally connected with all other thalamic nuclei), the RTN is ideally suited to control rhythmic thalamo-cortical activities[34,35,36,37]. Experimental manipulation of RTN low threshold Ca^{2+} conductances in the Wistar rat model has been found to induce ipsilateral dramatic changes in SW discharges[38]. These results are particularly interesting in view of the recent report of a blocking effect of ethosuximide (a well known antiabsence drug) on low threshold Ca^{2+} current in thalamic neurons[39]. Moreover, since the RTN is entirely made up of GABAergic neurons, GABAergic drugs are expected to exert an adverse effect on non convulsive generalized epilepsies by enhancing RTN-dependent synchronous discharges, which is indeed the case as reported in more detail below. It must be stressed, however, that identification of a key role of RTN as a control structure for SW generation does not necessarily imply that the genetically-determined defect responsible for non convulsive generalized seizures is to be found in this nucleus.

A sustained cortical hyperexcitability may in fact result in rhythmic paroxysmal discharges by impinging on thalamic oscillatory systems, which are then overdriven to a pathological rhythm which is eventually imposed to the entire thalamo-cortical complex. At present, the site and type of cellular dysfunction leading to genetically determined epileptogenesis cannot be satisfactorily defined. Many biochemical defects have been reported to affect neurotransmitters, receptors, membrane conductances, glycogen metabolism, etc., but it is unclear whether these have an etiologic significance or are merely secondary to the epileptic disorder.

To improve our understanding of the epileptogenic mechanism, the identification of the genes responsible for animal epilepsies is mandatory. This has only been possible for some mice mutants including tottering: the gene expression has been traced, leading to the demonstration of

34

a noradrenergic hyperfunction[11]. Attempts to characterize the specific mechanisms by which the excess of noradrenaline induces a hyperexcitable epileptogenic condition have been so far unsuccessful[40], but this is of course the approach which is expected to clarify the many questions which are still open.

Pharmacological studies in genetic animal models of epilepsy

Rats with petit-mal like epilepsy and tottering mice are ideal models to test antiepileptic drug efficacy due to the very high frequency of spontaneous seizures, which make it possible to carry out acute studies without unnatural seizure induction. Concerning Papio papio and mongolian gerbil, the possibility to elicit seizures will make them also theoretically suitable for acute as well as chronic studies. In the latter case, particular attention should be devoted to the reproducibility of stimulus characteristics over the time. Other problems may arise from fluctuations of stimulus-sensitivity as typically occur in Papio papio[25,28].

A comprehensive review of studies on the efficacy of the antiepileptic drugs commonly used in human epilepsies can be found in Löscher[2] and Löscher and Schmidt[3]. Relevant data to the animal models considered in the present review as compared with human epilepsies with generalized non convulsive seizures, are summarized in Table 1, modified from Löscher and Schmidt[3]. In rats with SW discharges, ED_{50} is the dose which decreases by 50% the duration of SW discharges in the 60 min period after drug administration, while in tottering mice ED is the dose which decreases the number of SW discharges. The three drugs, which are highly effective on human petit-mal and myoclonic epilepsies (valproic acid, ethosuximide and trimethadione) are also effective on animal models of nonconvulsive generalized seizures with particular regard to rats with SW. Benzodiazepines are also effective, while phenytoin and carbamazepine are ineffective or even aggravate the seizure, in agreement with clinical experience in human petit-mal. Phenobarbitone, which is effective in human epilepsies with myoclonic seizures, but not in genuine petit-mal, is active in the animal models but in rats with SW the effect is lost at higher doses (20 mg/kg). According to the profile of drug efficacy, rats with SW can be considered a good model for testing antiabsence

TABLE 1

EFFECTIVE DOSES OF ANTIEPILEPTIC DRUGS (MG/KG) ON EPILEPSIES WITH GENERALIZED NON CONVULSIVE SEIZURES (MODIFIED FROM LÖSCHER AND SCHMIDT[3])

	Rats with SW (ED_{50})	Tottering mice (ED)	Gerbils (ED_{50})	Papio papio (ED)	Clinical dose
Valproic acid	70		210	200	20–50
Ethosuximide	15	150	360	NE	15–20
Trimethadione	70			weakly	20–40
Diazepam	0.7	1.4	0.25	0.5–1	not used
Clonazepam			0.029	0.15	0.1–0.2
Phenytoin	NE	NE	NE	15–50	NE
Carbamazepine	NE		~40	40	NE
Phenobarbital	2	25	14	15	2–3
Primidone				NE	10–15

NE: not effective.

TABLE 2

EFFECTIVE DOSES (MG/KG) OF NEWLY DEVELOPED ANTIEPILEPTIC DRUGS

Drug	Rats with SW	Gerbils	Papio papio	Reference
Gabapentin		10–40 p.o.	NE	51
Progabide	AE	58	60	42,45,46
Vigabatrin	AE		450–950	43,46
THIP	AE	1	NE	44,45,46
β-Carboline-full agonists	0.5–4	?	0.5–1	46,49,52
β-Carboline-partial agonists	1–16	?	1	
β-Carboline-inverse agonists	AE			
Milacemide		NE	NE	53,54
2-AHP (excitatory aminoacid antagonist)			1 mM/kg	50
Flunarizine			2 i.v.	55
Ralitoline		10 p.o.	5 p.o.	56
Piperidino-pyridazine derivatives			40–50	57

AE: adverse effect; NE: not effective.

drugs, while baboons with photomyoclonic seizures, gerbils with myoclonic seizures and, to some extent, tottering mice seem to be more suitable to test drugs aimed to control epilepsies with a myoclonic component.

These considerations may provide a basis for evaluating the efficacy of newly developed antiepileptic drugs.

Starting from the hypothesis that impairment of GABAergic inhibition may be a pathophysiological epileptogenic mechanism[41], several GABAergic compounds have been proposed as potentially effective antiepileptic drugs. Among them vigabatrin (an irreversible inhibitor of the GABA degrading enzyme GABA-transaminase), progabide and THIP (two GABA receptor agonists) have been tested on rats with SW discharges, gerbils and Papio papio.

A protective effect against photomyoclonic seizures in Papio papio was produced by progabide[42], vigabatrin[43], but not by THIP[44]. Progabide and THIP were also found to be effective against myoclonic seizures in gerbils[45]. On the contrary, these compounds, like all other GABA-mimetic drugs tested so far, consistently aggravated the seizures in rats with SW[46], in agreement with the concept that an increase in GABAergic transmission may be involved in the pathogenesis of generalized SW discharges[47]. It must be noted, however, that benzodiazepine receptor agonists of the beta-carbolines group antagonize SW discharges in the same model[48], suggesting a complex involvement of the GABA-benzodiazepine receptors in the pathophysiology of non convulsive seizure with SW. In addition, the beta-carboline ZK 91296, a partial benzodiazepine receptor agonist, has been found to be effective in gerbils against both myoclonic and convulsive seizures[49].

Among the most recently developed drugs, an interesting class is represented by excitatory amino-acid antagonists such as 2-amino-7-phosphonoheptanoic acid (2-APH). This compound selectively antagonizes excitatory neurotransmission by acting on N-methyl-D-aspartate (NMDA) receptors. 2-APH 1 mM/kg i.v. has been shown to completely abolish the photomyoclonic response in Papio papio for 5 hours[50].

The effects of some other new drugs with different mechanisms of action are summarized in Table 2.

Conclusions

Experimental epilepsy has been studied in animal models seizures induced by chemical or physical means. The mechanisms of action of such epileptogenic agents have been further investigated at cellular or subcellular level in *in vitro* preparations of tissue slices or cell cultures. These approaches have considerably improved our understanding of the basic mechanisms of the epilepsies and have provided us with useful epilepsy models to be used in the search for new antiepileptic drugs. However, besides these advantages, some major drawbacks have to be taken into account. Animals in which seizures are induced by acute procedures, such as electroshock or pentylenetetrazol, are models for epileptic seizures rather than models of epilepsies. On the other hand, animals with recurrent seizures induced by chronic procedures, such as aluminum or cobalt foci, are, with few exceptions, suitable for partial but not for generalized epilepsies.

In any case, once the mechanism of action of a given epileptogenic agent is defined it is impossible to demonstrate its specific relevance for human epilepsies.

Genetic models of epilepsies provide us with a unique opportunity to study naturally occurring epilepsies and to test on them the effect of antiepileptic drugs.

The models which have been reviewed here resemble some forms of human idiopathic epilepsies. In particular, Wistar rats with non convulsive generalized epilepsies and, to a lesser extent, tottering mice can be proposed as a model of human petit mal, while photosensitive Papio papios reproduce some forms of human photogenic epilepsies. Less convincing is the proposed assimilation of 'xenogenic' seizures to some human myoclonic epilepsies, namely the juvenile myoclonic epilepsy[30].

In a strict sense, however, none of these models reproduce the complex of signs and symptoms which characterize the human syndromes. Indeed, this is not surprising in view of the considerable differences between man, baboon, gerbil, rat and mice in terms of the anatomo-physiological organization of the central nervous system. If one takes into account these differences, the similarities between human and animal forms look even closer than expected. In particular, in both humans and animals the inborn error which is responsible for epileptic seizures is expressed only after completion of some fundamental maturational steps in the central nervous system. Both in humans and animals recurrent epileptic seizures can be the only manifestation of the genetically-determined dysfunction. Both human and animal genetic epilepsies express themselves through a rather limited repertoire of seizures which shows superimposable electroclinical characteristics in different animal species.

Finally, human and animal epilepsies show superimposable profiles of sensitivity to antiepileptic drugs so that the specific efficacy of a given drug on a definite type of seizure in animals is a good predictor of its clinical efficacy on the same type of seizure in humans.

For these reasons, we believe that the importance of genetic animal models of epilepsies will be more and more recognized in pathophysiological and pharmacological studies on epilepsies leading to substantial advances which may eventually trace the way which links the molecular level to the clinical correlate.

Acknowledgements

The study was partially supported by the Paolo Zorzi Association for Neurosciences. We thank Maria Teresa Pasquali for her assistance in editing the text.

References

1 Commission on Classification and Terminology of the International League against Epilepsy (1989) Epilepsia 30: 389–399.

2 Löscher, W. (1984) Meth and Find Exptl. Clin. Pharmacol. 6: 531–547.

3 Löscher, W., Schmidt, D. (1988) Epilepsy Res. 2: 145–181.

4 Commission on Classification and Terminology of the International League against Epilepsy (1981) Epilepsia 22: 489–501.

5 Consroe, P., Edmonds, H.L. (1979) Fed. Proc. 38: 2397–2398.

6 Löscher, W., Meldrum, B.S. (1984) Fed. Proc. 43: 276–284

7 Green, M.C., Sidman, R.L. (1962) J. Hered. 53: 233–237.

8 Noebles, J.L., Sidman, R.L. (1979) Science 204: 1334–1336.

9 Noebels, J.L. (1979) Fed. Proc. 38: 2405–2410.

10 Levitt, P., Noebels, J.L. (1981) Proc. Natl. Acad. Sci. USA 78: 4630–4634.

11 Noebels, J.L. (1984) Nature 310: 409–411.

12 Seyfried, T.N., Glaser, G.N. (1985) Epilepsia 26: 143–150.

13 Chauvel, P., Louvel, J., Kurcewicz, I., Debono, M. (1980) In: Baumann, N. (ed.) Neurological Mutations Affecting Myelination. Elsevier, New York, pp. 513–516.

14 Vergnes, M., Marescaux, G., Micheletti, G., Reis, J., Depaulis, A., Rumbach, L., Warter, J.M. (1982) Neurosci. Lett. 33: 97–101.

15 Coenen, A.M.L., Van Luijtelaar, E.L.J.M. (1987) Epilepsy Res. 1: 297–301.

16 Thiessen, D.D., Lindzey, G., Friend, H.C. (1968) Psychon. Sci. 11: 227–228.

17 Goldblatt, D., Konow, A., Shoulson, I., Mac Math, T. (1971) Neurology (Minneap.) 21: 433–434.

18 Loskota, W.J., Lomax, P., Rich, S.T. (1974) Epilepsia 15: 109–119.

19 Majakowski, J., Donadio, M. (1984) Electroencephalogr. Clin. Neurophysiol. 57: 369–377.

20 Berry, K., Wisniewski, H.M., Schwartzbei, L. Baez, S. (1975) J. Neurol. Sci. 25: 75–92.

21 Schonfeld, A.R., Glick, S.D. (1979) Brain Res. 173: 147–151.

22 Donadio, M.F., Kozlowski, P.B., Kaplan, H., Wisniewski, H.M., Majkowski, J. (1982) Brain Res. 234: 263–273.

23 Majakowski, J. (1989) In: Beaumanoir, A., Gastaut, H., Naquet, R. (eds.) Reflex Seizures and Reflex Epilepsies. Editions Medicine Hygiene, Geneve, pp 19–24.

24 Killam, K.F., Naquet, R., Bert, J. (1966) Epilepsia 7: 215–219.

25 Balzamo, E., Bert, J., Menini, C.H., Naquet, R. (1975) Epilepsia 16: 269–276.

26 Killam, K.F., Killam, E.K., Naquet, R.J. (1967) Electroencephalogr. Clin. Neurophysiol. 22: 497–513.

27 Killam, K.F. (1969) Epilepsia 10: 229–238.

28 Wada, J.A., Terao, A., Booker, H.E. (1972) Neurology 22: 1272–1285.

29 Corcoran, M.E., Cain, D.P., Wada, J.A. (1979) Can. J. Neurol. Sci. 6: 129–131.

30 Delgado Escueta, A.V., White, R., Greenberg, W.A., Treiman, L.J. (1986) In: Delgado Escueta, A.V., Ward, Jr. A.A., Woodbury, D.M., Porter, R.J. (eds.) Advances in Neurology vol. 44, Basic Mechanisms of the Epilepsies. Raven Press, New York, pp. 77–95.

31 Fisher-Williams, M., Poncet, M., Riche, D., Naquet, R. (1968) Electroencephalogr. Clin. Neurophysiol. 25: 557–569.

32 Menini, C., Silva-Barret, C. (1989) In: Beaumanoir, A., Gastaut, H., Naquet, R. (eds.) Reflex Seizures and Reflex Epilepsies, Editions Medicine & Hygiene, Geneve, pp. 39–48.

33 Silva-Barrat, G., Menini, G., Bryere, P., Naquet, R. (1986) Electroencephalogr. Clin. Neurophysiol. 64: 455–468.

34 Avanzini, G., de Curtis, M., Panzica, F., Spreafico, R. (1989) J. Physiol. 416: 111–122.

35 Steriade, M., Deschenes, M. (1984) Brain Res. Rev. 8: 1–63.

36 Mulle, C., Madariaga, A., Deschenes, M. (1986) J. Neurosci 6: 2134–2145.

37 de Curtis, M., Spreafico, R., Avanzini, G. (1989) Neuroscience 33: 275–283.

38 Avanzini, G., Marescaux, C., Spreafico, R., Vergnes, M. (in preparation).

39 Coulter, D.A., Huguenard, J.R., Prince, D.A. (1989) Ann. Neurol. 25: 582–593.

40 Kostopoulus, G., Psarropoulou, G., Haas, H.L. (1988) Exp. Brain Res. 72: 45–50.

41 Schwartzkroin, P.A., Prince, D.A. (1980) Brain Res. 183: 61–76.

42 Cepeda, C., Worms, P., Lloyd, K.G., Naquet, R. (1982) Epilepsia 23: 463–470.

43 Meldrum, B.S. (1978) In: W.A. Cobb, Wan Duijn, H. (eds) Contemporary Clinical Neurophysiology. Elsevier, Amsterdam, pp 317–322.

44 Meldrum, B., Horton, R. (1980) Eur. J. Pharmacol. 61: 231–237.

38

45 Löscher, W., Frey, H.H. (1984) Arzneim. Forsch. 34: 1484–1488.

46 Marescaux, C., Vergnes, M., Depaulis, A., Micheletti, G., Walter, J.M. (1990) In: Avanzini, G., Fariello, R., Heinemann, U., Engel, J. (eds.) Advances in the Neurobiology of Epilepsy. Neurotransmitters. Demos Publication, New York, pp 453–465.

47 Gloor, P., Fariello, R.G. (1988) Trends in Neurosciences 11: 63–68.

48 Marescaux, C., Vergnes, M., Jensen, L., Petersen, E., Depaulis, A., Micheletti, G., Warter, J.M. (1987) Brain Res. Bull. 19: 327–335.

49 Löscher, W., Schneider, H., Kehr, W. (1985) Eur. J. Pharmacol. 114: 261–266.

50 Meldrum, B.S., Croucher, M.J., Badman, G., Collins, J.F. (1983) Neurosci. Lett. 33: 101–104.

51 Bartoszyk, G.D., Meyerson, N., Reimann, W., Satzinger, G., von Hodenberg, A. (1986a) In: Meldrum, B.S., Porter, R.J. (eds.) Current Problems in Epilepsy vol. 4, New Anticonvulsant Drugs. John Libbey, London, pp 147–163.

52 Meldrum, B. (1984) Neuropharmacol. 23: 845–846.

53 Löscher, W. (1985) J. Pharmacol. Exp. Ther. 233: 204–213.

54 Roba, J., Cavalier, R., Cordi, A., Gorissen, H., Herin, M., Janssens de Varebeke, P., Onkelinx, C., Remacle, M., van Dorsser, W. (1986) In: Meldrum, B.S., Porter, R.J. (eds.) Current Problems in Epilepsy vol. 4, New Anticonvulsant Drugs. John Libbey, London, pp 179–190.

55 De Sarro, G.B., Nisticò, G., Meldrum, B.S. (1986) Neuropharmacol. 25: 695–7O1.

56 Bartoszyk, G.D., Dooley, D.J., Fritschi, E., Satzinger, G. (1986b) In: Meldrum, B.S., Porter, R.J. (eds) Current Problems in Epilepsy vol. 4, New Anticonvulsant Drugs. John Libbey, London, pp 309–311.

57 Chambon, J.P., Hallot, A., Biziere, K. (1986) In: Meldrum, B.S., Porter, R.J. (eds.) Current Problems in Epilepsy vol. 4, New Anticonvulsant Drugs. John Libbey, London, 313–316.

New Antiepileptic Drugs (Epilepsy Res. Suppl. 3)
Editors: F. Pisani, E. Perucca, G. Avanzini, A. Richens
© 1991 Elsevier Science Publishers B.V. (Biomedical Division)

CHAPTER 5

Excitatory amino acids in epilepsy and novel anti-epileptic drugs

Astrid Chapman, Brian Meldrum

Department of Neurology, Institute of Psychiatry, De Crespigny Park, London SE5 8AF, UK

Introduction

The excitatory neurotransmitters glutamate and aspartate are critically involved in the initiation and spread of seizure activity. Our knowledge of these processes has developed extensively over the last ten years but is still defective in many respects. In parallel with developments in our basic understanding there have been important advances in terms of therapeutic approaches based on manipulation of excitatory neurotransmission. This has reached the stage of initial clinical evaluation for some excitatory amino acid antagonists. We shall here review firstly evidence relating to the involvement of excitatory amino acids in epileptic phenomena and secondly the current status of excitatory amino acid antagonists as potential therapeutic agents in epilepsy.

Excitatory amino acids and burst discharges

Intracellular recordings both *in vivo* and *in vitro* have revealed a pattern of neuronal discharge that is considered to typify epileptic activity. This consists of a large membrane depolarisation resembling a giant EPSP and referred to as a 'paroxysmal depolarising shift' associated with a rapid burst of action potentials. This pattern of activity is seen in neurones in a spontaneous or chemically-induced cortical focus and corresponds to an interictal spike recorded with gross electrodes[1]. The pattern is also seen in generalised seizures associated with repetitive spikes or spikes and waves and in *in vitro* preparations such as hippocampal slices exposed to ionic conditions or convulsant agents that favour spontaneous or electrically-induced synchronous discharges.

Following the description of the three main types of postsynaptic glutamate receptor (kainate, AMPA/quisqualate and N-methyl-D-aspartate, NMDA) it was shown that the iontophoretic application of NMDA (or other agents acting on the NMDA receptor such as quinolinate or homcysteate) to normal neurones in the hippocampus, striatum or neocortex could induce a pattern of burst discharge, with a 'paroxysmal depolarising shift' that was closely similar to that seen in an epileptic focus[2]. The voltage-dependent block of the NMDA receptor by Mg^{2+} probably explains the paroxysmal depolarising shift induced by NMDA application. Spontaneous burst firing occurring *in vivo* and *in vitro* also commonly depends on a powerful, synchronous excitatory input. In hippocampal slices burst firing can be induced in CA3 or CA1 pyramidal neurones by ionic alterations, low $[Mg^{2+}]$, low $[Ca^{2+}]$ or high

[K$^+$] and by various convulsants including kainate and GABA receptor antagonists such as bicuculline and picrotoxin. Under some circumstances (e.g. high [Mg^{2+}]) the burst firing can be completely abolished by the bath application of rather low concentrations of NMDA receptor antagonists (2-amino-5-phosphonovalerate, 10 µmol). Under other circumstances (e.g. kainate or high [K$^+$]) non-NMDA antagonists such as the quinoxalinedione (CNQX) can block the burst discharges[3,4]. In the presence of GABA antagonists (penicillin, bicuculline) the initial component of the burst is dependent on activation of non-NMDA receptors and the later components on NMDA receptors. Thus the effect of NMDA antagonists is to shorten the duration of the bursts[5,6]. Hippocampal slices derived from rats with prior kainate lesions of the CA3 neurones show burst firing with variable pharmacological responsiveness (sometimes suppressed by NMDA - sometimes by non-NMDA antagonists)[7].

Abnormalities of excitatory neurotransmission and the pathogenesis of epilepsy

The pathogenesis of epilepsy in terms of molecular and cellular mechanisms is not well understood. Two major hypotheses concern inhibitory and excitatory neurotransmission. One supposes that there is some failure in GABAergic inhibitory systems. The data concerning GABA antagonists in slice preparations indicate that such a failure could lead to burst firing. Studies in cortical foci induced by alumina in the monkey have suggested that loss of GABAergic synapses may be an important factor[8,9]. In human epileptic foci immunocytochemical studies have not shown a selective loss of GABAergic neurones or terminals[10]. The alternative hypothesis that enhanced excitatory function is involved in epileptogenesis has recently gained support from studies in experimental models of epilepsy and in human material obtained post-mortem or through surgical intervention. Morphological studies have defined a pattern of synaptic reorganisation ('sprouting') that could provide an excitatory feedback within the hippocampus. An abnormal pattern of regrowth of the mossy fibre system to supply the inner third of the molecular layer in the dentate gyrus was initially observed following kainic acid lesioning of the CA3 region of the hippocampus[11]. Subsequently a similar pattern of abnormal growth was observed in kindled rats using Timm's stain[12] or autoradiography with [^{3}H]kainate[13]. Studies with either Timm's stain[14] or with immuno-cytochemistry for dynorphin[15,16] in human hippocampi obtained by anterior temporal lobectomy in patients with intractable complex partial seizures also reveal abnormal sprouting of the mossy fibres in the inner part of the molecular layer. The contribution that this morphological change makes to epileptogenesis is not yet defined. If it is providing a powerful direct excitatory feedback to granule cells it could contribute to the kindled seizure state and to the intractability of some cases of temporal lobe epilepsy.

There is also evidence for changes in excitatory amino acid receptors associated with epilepsy. This evidence concerns both receptor binding as revealed in autoradiography, and receptor function as revealed in electrophysiological studies. In animal models of epilepsy receptor autoradiography has tended not to provide consistent evidence for changes in glutamate receptor number or affinity either in genetic models[17] or in kindling[18,19,20,21,22]. However studies utilising anterior temporal lobectomy specimens from patients with intractable seizures have shown changes which although not quantitatively consistent between studies do show some consistent trends. Comparing 5 operative samples with 6 autopsy controls Geddes et al.[23] found a loss of [^{3}H]kainate and NMDA-sensitive L-[^{3}H]glutamate binding in CA3 and CA1. Moving from the sclerotic region to the hippocampal cortex, binding increased to above normal such that there was a two-fold increase in L-[³H]glutamate binding and some increase in [³H]kainate binding in the parahippocampal gyrus. In a similar study McDonald et al.[24] found binding to phencyclidine (PCP) receptors to be consistently reduced in epileptic hippocampi compared with autopsy controls. Binding to the NMDA recognition site and its glycine modulatory site was increased in area CA1 and the dentate gyrus, whereas binding to the quisqualate receptor was unchanged in these regions. Both studies suggest that as cells are lost in the hippocampus there is a relative increase in NMDA receptor density.

Electrophysiological studies in kindled rats have provided rather strong evidence for an increased sensitivity of NMDA receptors.

An early study of hippocampal slices from hippocampal-kindled rats showed a relative increase in the magnitude of NMDA receptor mediated changes in $[Ca^{2+}]$o at sites distal to the cell bodies of CA1 pyramidal neurones[25]. Subsequently in amygdala-kindled rats it was shown that perforant path stimulation induced NMDA-receptor mediated post-synaptic potentials that were not present in non-kindled animals[26]. In hippocampal wedge preparations from kindled rats the depolarising action of NMDA is enhanced in CA3 (but not in CA1)[27].

Kindling is also associated with a decrease in the cortical content of glutamine[28] and an increase in the synaptic release of glutamate[29].

Electrophysiology in man
Although definitive proof is not available it is clearly possible that an abnormal pattern of responsiveness of NMDA receptors is acquired during the ontogenesis of many different forms of epilepsy. This would include an abnormal distribution of NMDA-responsive ionophores in hip-

pocampus and neocortex. By facilitating the induction and maintenance of burst firing these receptors could contribute significantly to the refractory nature of focal neocortical or temporal lobe seizures. Louvel and Pumain[30] have studied focal epileptic and control cortex (tumour, non-epileptic) in vitro using ion-sensitive electrodes to measure $[Ca^{2+}]$o. In the epileptic cortex the response to NMDA is enhanced showing a much wider laminar distribution of peak sensitivity.

Mechanisms for decreasing excitatory neurotransmission

Regardless of the importance of abnormal excitatory neurotransmission in the initiation of seizures it is clear that reducing excitatory neurotransmission could provide a rational approach to anticonvulsant therapy, so long as normal function was not also impaired.

There are many potential pharmacological approaches to reducing excitatory neurotransmission. They are summarised in Table 1. Most of the synaptically released glutamate derives from glutamine (released from glial cells following the action of glutamine synthetase). Various inhibitors are known to act on glutaminase. Given systemically DON and azaserine have weak but measurable anticonvulsant actions[31]. Given focally into the inferior colliculus L-canaline potently suppresses audiogenic seizures in rats[32].

The release of excitatory amino acids is under a wide variety of presynaptic controls. GABA and GABA_B agonists such as baclofen potently reduce stimulated release of glutamate in in vitro preparations. However GABA_B agonists are not antiepileptic agents, probably because they also (more potently) act presynaptically to decrease GABA release. Adenosine and various stable adenosine analogues act presynaptically to decrease glutamate release. Various antiepileptic effects are described in in vitro preparations[33] but effects

TABLE 1

PHARMACOLOGICAL APPROACHES TO DECREASE GLUTAMATERGIC NEUROTRANSMISSION

1.	Decreased glutamate synthesis
	Glutaminase inhibitors (azaserine, DON)
2.	Presynaptic actions on glutamate release
	a) adenosine analogues
	b) GABA$_B$ agonists
	c) Other (lamotrigine, phenytoin)
3.	Post-synaptic receptor antagonists
	a) Non-selective
	b) Selective NMDA antagonists
	c) Kainate/quisqualate antagonists
	d) Metabotropic receptor antagonists
4.	Enhanced glutamate uptake
5.	Long-term down regulation of EAA system

in vivo seem very limited.

Some established anticonvulsant drugs (such as phenytoin and benzodiazepines) decrease the stimulated release of excitatory amino acids in *in vitro* preparations, as does the novel anticonvulsant lamotrigine[34].

Potent blockade of excitatory transmission can be produced by a wide variety of drugs acting at post-synaptic sites. Compounds may act non-selectively on all receptor subtypes, or with varing degrees of selectivity. They may also act as competitive antagonists at the agonist recognition site or as non-competitive antagonists at various regulatory or allosteric sites. The most convincing body of data concerns the anticonvulsant action of selective NMDA receptor antagonists. Compounds with broad spectrum antagonism also show some antiepileptic action. Compounds with selective kainate or quisqualate antagonism are only feebly anti-epileptic in most standard test systems.

NMDA receptor antagonists as anticonvulsant agents

The principal categories of NMDA antagonist are listed in Table 2. The molecular structures of the principal competitive NMDA antagonists are shown in Figure 1. Potent anticonvulsant effects have been demonstrated for competitive NMDA

TABLE 2

SELECTIVE NMDA ANTAGONISTS

1.	Competitive antagonists
	AP5, AP7, CPP, CPP-ene, CGS 19755, CGP 37849, CGP 39551
2.	Glycine antagonists
	HA 966, 7-Cl-KYN, 5,7-Cl-KYN
3.	Non-competitive antagonists
	MK-801, PCP, TCP, dextrorphan
4.	Polyamine site antagonists
	Ifenprodil, SL 82,0715

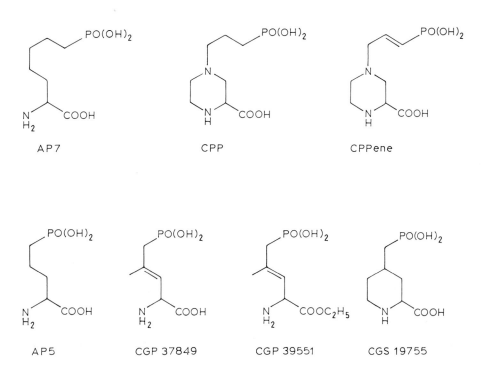

Fig. 1. Molecular structures of competitive NMDA antagonists. CPP and its unsaturated analogue D-CPPene are analogues of AP7. CGS 19755, CGP 37849 and CGP 39551 are analogues of AP5.

antagonists and for non-competitive antagonists of the MK-801/PCP type. Glycine site antagonists have some anticonvulsant action but compounds so far tested fall short of the potency of the first two types. NMDA antagonists of the 'polyamine antagonist' type are feebly or not at all anticonvulsant.

Competitive NMDA antagonists in rodent models

Competitive NMDA antagonists when given intracerebroventricularly in DBA/2 mice are more potent anticonvulsants than any of the established antiepileptic compounds (see Table 3). The relative potencies of different antagonists correspond fairly closely to their relative potencies at displacing [3H]CPP from binding sites in brain membrane preparations.

The competitive NMDA antagonists are relatively less potent compared to established antiepileptic drugs when given intravenously or intraperitoneally, suggesting that they penetrate less effectively to relevant brain sites than benzodiazepines, phenytoin, etc. Among the different competitive antagonists the relative potencies are comparable by either route suggesting that there is no noticeable difference between the different competitive antagonists in terms of their access to the brain from the blood.

The anticonvulsant efficacy of the competitive antagonists when administered orally was initially overlooked because of the use of low doses and short predosing times. However with testing at 3–4 hours the competitive antagonists do show po-

44

TABLE 3

NMDA ANTAGONISTS AND STANDARD ANTICONVULSANT DRUGS: POTENCY IN DBA/2 MICE

	ED_{50} Clonic Seizures (DBA/2 mice)		
	i.c.v. (nmol)	i.p. (μmol/kg)	p.o. (μmol/kg)
D(–)CPPene	0.008	1.54	40.2
D(–)CPP	0.022	2.75	65.8
CGS 19755	0.02	1.78	32.2
CGP 37849	0.04	3.27	33.9
CGP 39551	0.21	19.4	28.1
D(–)AP7	1.80	40.0	464.0
D,L(±)AP5	15.0	320	ND
DIAZEPAM	11	0.4	ND
PHENYTOIN	300	9.1	ND
PHENOBARBITONE	320	9.9	ND
VALPROATE	6000	1250	ND

tent anticonvulsant activity.

By this route one compound CGP 39551, the ethyl ester of CGP 37849, shows unexpectedly high oral activity.

Competitive NMDA antagonists in photosensitive baboons

Baboons, Papio papio, from the Casamance region of Senegal naturally show a syndrome of photosensitive epilepsy, which is manifest during stroboscopic stimulation as myoclonus. The myoclonus initially involves the muscles of the eye-lids, subsequently those of the head and neck and then the trunk and limbs. It may then become rhythmic and self-sustaining so that it continues beyond the end of photic stimulation, or progresses to a tonic clonic seizure. This provides a means of scoring responses to a standardised stimulation, which can be repeated at intervals following drug administration. By this means standard and novel anticonvulsant drugs have been tested in Papio papio[38–40]. Effective and toxic plasma anticonvulsant drug concentrations in the baboon are broadly similar to those established in the clinic.

TABLE 4

SUPPRESSION OF PHOTICALLY-INDUCED MYOCLONUS BY NMDA ANTAGONISTS IN PAPIO PAPIO

Antagonist	ED_{100}	(mg/kg)
	i.v	oral
D,L AP7	200	NT
D,L AP5	600	NT
D (−) CPP	NT	32
D (−) CPPene	8	24
CGP 37849	40	40
CGP 39551	≥160	40
MK-801	0.034	NT

The intravenous administration of competitive NMDA antagonists leads to a prompt (onset 5–15 min, peak 1–2 h) suppression of myoclonic responses to photic stimulation[39].

Table 4 summarises the relative potencies of the principal competitive NMDA antagonists.

The demonstration that CPP, CPPene, CGP 37849 and CGP 39551 are orally active in rodents[41] lead to the oral testing of these compounds in baboons. All 4 compounds show a similar time course of action with onset after 3–4 hours, and a prolonged duration of anticonvulsant effect. As

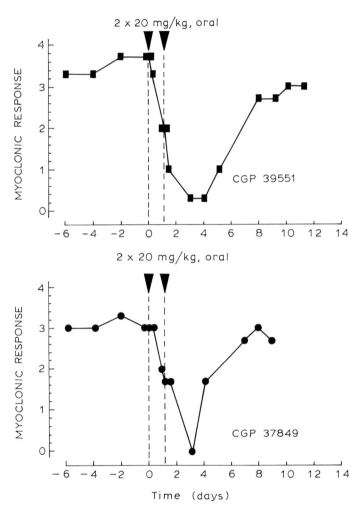

Fig. 2. Antiepileptic effect of competitive NMDA antagonists given orally in photosensitive baboons, Papio papio. Baboons were stimulated with a stroboscope at 25 Hz for up to 5 min and the myoclonic response scored as follows: 0 = none, 1 = myoclonus of eyelids, 2 = myoclonus of muscles of face and neck, 3 = myoclonus of trunk and limb muscles, 4 = myoclonus continuing after cessation of photic stimulation. Each point represents the mean score for 3 highly photosensitive baboons tested in the week before or the 2 weeks after the oral administration on 2 consecutive days of either CGP 39551 or CGP 37849.

in the rodents CGP 39551 is notably less potent than 37849 when given intravenously, but as potent or more potent when given orally.

Single daily doses have a cumulative effect such that a sustained anticonvulsant action is achieved after giving 20 mg/kg orally on two successive days (see Figure 2). Motor side effects (nystagmus, ataxia, muscle weakness or abnormal limb movements) were not observed after fully effective oral doses of CPP, CPPene, CGP 37849 and CGP 39551.

Non-competitive NMDA antagonists

PCP/MK-801 site

When tested against audiogenic seizures in DBA/2 mice or against electroshock seizures in mice or rats the dissociative anaesthetics phencyclidine and ketamine, MK-801 and some related compounds such as dextrorphan show anticonvulsant activity that correlates with their potency as non-competitive NMDA antagonists[35,42]. However these compounds also induce abnormal movements of the phencyclidine type (stereotypies, head weaving, ataxia) at doses at or below the anticonvulsant dose[43]. Similar results are obtained in the photosensitive baboon. Thus MK-801 is exceptionally potent in terms of suppression of myoclonic responses (see Table 4), but a fully effective dose (0.034 mg/kg) produces phencyclidine like side effects with loss of aggressivity, abnormal movements and impaired sensorium.

Glycine site

NMDA antagonists acting on the glycine site have some anticonvulsant activity. HA 966 is weakly active as an anticonvulsant when given systemically. 7-Chlorokynurenic acid has a modest short-lasting anticonvulsant action when given intracerebroventricularly to DBA/2 mice[44]. The development of more potent glycine antagonists with appropriate pharmacokinetics is required

before we can appraise their potential as anticonvulsant agents.

Side effects of NMDA antagonists

All NMDA antagonists produce some motor side effects. Competitive NMDA antagonists produce motor weakness and ataxia, which in rodents is prominently manifest as hindlimb flaccidity. However rotarod testing in DBA/2 mice reveals that the therapeutic margin for CPP, CPPene, CGP 37849 and CGP 39551 is as good as or better than that of several established anticonvulsant drugs[41,45]. This is not the case however for the non-competitive antagonists of the PCP-MK-801 type.

There may be other important effects of NMDA antagonists on cognitive functioning, in particular learning and memory may be impaired because of the important role played by the NMDA receptor system in synaptic plasticity (as seen in experiments on long term potentiation). NMDA receptor blockade has been shown to impair the acquisition of certain spatial learning tasks in the rodent[46,47]. Synaptogenesis may also be modified by NMDA receptor antagonists at certain crucial phases in development[48].

Assessment of the range and severity of motor and cognitive side effects during phase I testing in man will be crucial to the future prospects of NMDA antagonists as antiepileptic agents.

Future prospects for NMDA antagonists in epilepsy. Which will be the optimal antagonist?

Of the NMDA antagonists currently available it is clear that the competitive antagonists offer the most immediate prospect for testing in man. Although the different competitive antagonists appear to be broadly similar in terms of anticonvulsant profile and side effects they are not identical. They differ in terms of potency and kinetics at the receptor site and they may also differ in terms of action at whatever NMDA receptor subtypes emerge in the next few years. (Such differences are the most probable explanation for differences in the therapeutic ratios of different competitive antagonists). It is also not possible to forecast which syndromes of epilepsy will be most responsive to NMDA receptor antagonists. However if enhanced sensitivity to NMDA plays as important a role in the establishment of refractory epilepsy as recent observations suggest, then trials in patients with refractory complex partial seizures may well prove successful.

Acknowledgements

We thank the British Epilepsy Research Foundation and the Medical Research Council for financial support. We thank Dr S. Patel, and Ms. J. Graham for their collaboration in the anticonvulsant drug studies.

References

1 Ayala, G.F., Matsumoto, H., Gumnit, R.J. (1970) J. Neurophysiol. 33: 73–85.
2 Herrling, P.L., Morris, R., Salt, T.E. (1983) J. Physiol. 339: 207–222.
3 Neuman, R., Cherubini, E., Ben-Ari, Y. (1988) Brain Res. 459: 265–274.
4 Ben-Ari, Y., Gho, M. (1988) J. Physiol. 404: 365–384.
5 Dingledine, R., Hynes, M.A., King, G.L. (1986) J. Physiol. 380: 175–189.
6 Schneiderman, J.H., MacDonald, J.F. (1989) Neuroscience 31: 593–603.
7 Simpson, L.H., Wheal, H.V., Williamson, R. (1990) Can. J. Physiol. Pharmacol. (In press).
8 Ribak, C.E., Joubran, C., Kesslak, J.P., Bakay, R.A.E. (1989) Epilepsy Res. 4: 126–138.
9 Ribak, C.E., Harris, A.B., Vaughn, J.E., Roberts, E. (1979) Science 205: 211–214.
10 Babb, T.L., Pretorius, J.K., Kupfer, W.R., Crandall, P.H. (1989) J. Neurosci. 9: 2562–2574.
11 Nadler, J.V., Martin, D., Bowe, M.A., Morrisett, R.A., McNamara, J.O. (1990) In: Ben-Ari, Y. (ed.) Excitatory Amino Acids and Neuronal Plasticity. Plenum Press, New York, pp. 1–10.
12 Sutula, T., He, X.X., Cavazos, J., Scott, G. (1988) Science 239, 1147–1150.
13 Represa, A., Le Gall La Salle, G. and Ben-Ari, Y. (1989) Neurosci. Lett. 99: 345–350.
14 Sutula, T., Cascino, G., Cavazos, J., Parada, I., Ramirez, L. (1989) Ann. Neurol. 26: 321–330.
15 de Lanerolle, N.C., Spencer, D.D. (1990). (Unpublished).
16 Houser, C.R., Miyashiro, J.E., Swartz, B.E., Walsh, G.O., Rich, J.R., Delgado-Escueta, A.V. (1990) J. Neurosci. 10: 267–282.
17 Geddes, J.W., Cahan, L.D., Cooper, S.M., Kim, R.C., Choi, B.H., Cotman, C.W. (1990) Exp. Neurol. (In press).
18 Savage, D.D., Nadler, J.V., McNamara, J.O. (1984) Brain Res 323: 128–131.
19 Savage, D.D., Werling, L.L., Nadler, J.V., McNamara, J.O. (1982) Eur. J. Pharmacol. 85: 255–256.
20 Sircar, R., Ludvig, N., Zukin, S.R., Moshe, S.L. (1987) Eur. J. Pharmacol. 141: 167–168.
21 Jones, S.M., Johnson, K.M. (1989) Exp. Neurol. 106: 52–60.
22 Okazaki, M.M., McNamara, J.O., Nadler, J.V. (1989) Brain Res. 482: 359–363.
23 Geddes, J.W., Cooper, S.M., Cotman, C.W., Patel, S., Meldrum, B.S. (1989) Neuroscience 32: 39–47.
24 McDonald, J.W., Garofalo, E.A., Hood, T., Sackellares, J.C., Gilman, S., McKeever, P.E., Troncoso, J.C., Johnston, M.V. (1990) (In press).
25 Wadman, W.G., Heinemann, U. (1985) In: M. Kessler et al. (eds.) Physiology and Medicine. Springer Verlag, Berlin, pp. 221–228.
26 Mody, I., Stanton, P.K., Heinemann, U. (1988) J. Neurophysiol. 59: 1033–1053.
27 Nadler, J.V., Perry, B.W., Cotman, C.W. (1980) Brain Res. 182: 1–9.
28 Leach, M.J., Miller, A.A., O'Donnell, R.A., Webster, R.A. (1983) J Neurochem. 41: 1492–1494.
29 Jarvie, P.A., Logan, T.C., Geula, C., Slevin, J.T. (1990) Brain Res. 508: 188–193.
30 Louvel, J., Pumain, R. (1990) In: Engel, J. (ed.) Neurotransmitters, Seizures and Epilepsy IV. Raven Press, New York.
31 Chung, S.H., Johnson, M.S. (1984) Proc. Roy. Soc. Lond. B. 221: 145–168.
32 Faingold, C.L., Meldrum, B.S. (1990) In: Avoli, M., Gloor, P., Kostopoulos, G., Naquet, R. (eds.) Generalized Epilepsy. Neurobiological Approaches. Birkhauser, Boston.

48

33 Dragunow, M. (1988) Progr. Neurobiol. 31: 85–108.

34 Miller, A.A., Sawyer, D.A., Roth, B., Peck, A.W., Leach, M.J., Wheatley, P.L., Parsons, D.N., Morgan, R.J.I. (1986) In: Meldrum, B.S., Porter, R.J. (eds.) New Anticonvulsant Drugs. John Libbey, London, pp. 165–177.

35 Chapman, A.G., Meldrum, B.S. (1989) Eur. J. Pharmacol. 166: 201–211.

36 Chapman, A.G., Graham, J.L, Patel, S., Meldrum, B.S. (1990) Unpublished.

37 Fagg, G.E., Olpe, H.-R., Pozza, M.F., Baud, J., Steinmann, M., Schmutz, M., Portet, C., Baumann, P., Thedinga, K., Bittiger, H., Allgeier, H. Heckendorn, R., Angst, C., Brundish, D., Dingwall, J.G. (1990) Br. J. Pharmacol. 99: 791–797.

38 Meldrum, B.S., Horton, R.W., Toseland, P.A. (1975) Arch. Neurol. 32: 289–294.

39 Meldrum, B.S., Croucher, M.J., Badman, G., Collins, J.F. (1983) Neurosci. Lett. 39: 101–104.

40 Meldrum, B. (1984) Epilepsia 25: S140– S149.

41 Chapman, A.G., Graham, J., Meldrum, B.S. (1990) Eur. J. Pharmacol. 178: 97–99.

42 Nevins, M.E., Arnolde, S.M. (1989) Brain Res. 503: 1–4.

43 Koek, W., Colpaert, F.C. (1990) J. Pharmacol. Exp. Ther. 252: 349–357.

44 Singh, L., Oles, R.J., Tricklebank, M.D. (1990) Br. J. Pharmacol. 99: 285–288.

45 Patel, S,. Chapman, A.G., Graham, J.L., Meldrum, B.S., Frey, P. (1990) Epilepsy Res.

46 Morris, R.G.M., Anderson, E., Lynch, G.S., Baudry, M. (1986) Nature 319: 774–776.

47 Morris, R.G.M. (1989) J. Neurosci. 9: 3040–3057.

48 Bear, M.F., Kleinschmidt, A., Gu, Q., Singer, W. (1990) J. Neurosci. 10: 909–925.

New Antiepileptic Drugs (Epilepsy Res. Suppl. 3)
Editors: F. Pisani, E. Perucca, G. Avanzini, A. Richens
© 1991 Elsevier Science Publishers B.V. (Biomedical Division)

CHAPTER 6

Anticonvulsant activity of some calcium antagonists in genetically epilepsy prone rats

Giovambattista De Sarro, Giuseppe R. Trimarchi, Francesco Federico[1],
Angela De Sarro

*Institute of Pharmacology, School of Medicine, University of Messina and [1]Institute of Neurology, School of Medicine,
University of Bari, Italy*

Introduction

Epileptiform bursts are often associated with influx of calcium ions into nerve cells[1] and a decrease in the extracellular concentration of calcium precedes the onset of seizures in many experimental models of epilepsy[2]. Selective antagonists of calcium influx possess anticonvulsant properties in various animal models of experimental epilepsy[3-5] and in humans[6]. Several classes of calcium antagonists have been identified: A) those selective for slow Ca^{++} channels: *I*) the phenylalkylamines (verapamil, methoxyverapamil); *II*) the dihydropyridines (nicardipidine, lacidipine, nifedipine, nimodipine, nisoldipine, nitrendipine); *III*) the benzothiazepines (diltiazem); B) those non-selective for slow Ca^{++} channels: *IV*) the diphenylalkylamines (cinnarizine, flunarizine); *V*) the prenylamine-like group; *VI*) others (perhexiline)[7].

Binding studies have identified three distinct allosterically linked binding sites for the phenylalkylamine, dihydropyridine and benzothiazepine classes of calcium antagonists[8], whilst diphenylalkylamines utilize another binding site[9]. In the present study the behavioural and anticonvulsant effects of several compounds acting by various mechanisms on calcium-channels or affecting intracellular calcium concentrations were studied after systemic (intraperitoneal, i.p., and intracerebroventricular, i.c.v.) administration in genetically epilepsy prone rats strain-9 (GEPRs). These animals are genetically susceptible to sound-induced seizures[10].

Material and methods

Animals

GEPRs (a strain derived from Sprague-Dawley rats) were obtained from Dr B.S. Meldrum breeders, London University. The rats were housed three or four per cage in stable conditions of humidity (60 ± 5%) and temperature (21 ± 2°C) and allowed free access to food and water until the time of the experiments. Animals were maintained on a 12 h light and 12 h dark cycle.

Seizures were induced in GEPRs, 180–260 g, 10–18 weeks old, male or female by exposing them to a mixed frequency sound of 109 dB intensity under a hemispheric plexiglas dome (58 cm diameter). Auditory stimulation (109 dB) was applied for 60 sec or until the onset of convulsions occurred. Seizure response (S.R.) was assessed as previously reported by Jobe[10]. The maximum response was recorded for each animal.

Surgical procedures

I.c.v. injections were performed by means of a stainless guide cannula (22 gauge) chronically implanted, under chloral hydrate anaesthesia, into the left or right lateral ventricle according to atlas coordinates of Paxinos and Watson[11]. The hydrosoluble diltiazem, HA 1004, nicardipine and verapamil, belonging to different classes of calcium antagonists, were dissolved in sodium phosphate buffer 67 mM, pH 7.2–7.4, and 10 μl (at a rate of 2 μl/min) of vehicle (buffer) or drug solution was injected intraventricularly (i.c.v.) into the animals using an injector cannula connected through a Teflon tube to a Hamilton microsyringe (type 701N). For systemic injections, diltiazem, nicardipine, HA 1004 and verapamil were dissolved in sterile saline. All the other compounds were given i.p. (0.4 ml/10 g of body weight) as a freshly ultrasonicated suspension in 50% dimethylsulfoxide (DMSO) and 50% saline.

In order to avoid the light sensitivity of some of these compounds, weighing and handling were carried out under light from sodium vapour lamps and the substances were protected from light during the experiments.

Statistical analysis

The effects of treatment were analyzed by non-parametric methods. A Kruskall-Wallis analysis of variance was first carried out and if this was significant a Mann-Whitney U-test was used to compare control and treated groups.

The percentage of animals exhibiting the tonic extension (S.R. = 4–5) or clonic phase (S.R. = 2–3) of the audiogenic seizure was determined for each dose of calcium antagonist administered and these values were plotted against corresponding doses for calculation of ED_{50} values (with 95% confidence limits). The ED_{50} values for each compound and each phase of seizure response were estimated using the method of probit analysis[12].

Drugs

The sources of the drugs used were: cinnarizine and flunarizine (Janssen Farmaceutici, Roma, Italy), nicardipine (Sigma, St. Louis MO, U.S.A.), diltiazem (Sigma-Tau Res. Labs. Pomezia, Italy); verapamil and D-600 (methoxyverapamil) (Knoll Pharmaceutical Company, Ludwigshafen, West Germany) HA 1004 (N-(2-guanidino-ethyl-5-isoquinolinesulfonamide, Asahi Chemical Industry, Tokyo, Japan), nifedipine, nimodipine, nitrendipine, nisoldipine and Bay K 8644 (1,4-dihydro-2,6-dimethyl-3-nitro-4-(2-trifluorophenyl)pyridine-5-carboxylic acid) (Bayer Res. Labs. Milan, Italy), lacidipine (Glaxo Res. Labs. Verona, Italy), perhexiline maleate (Sigma, St. Louis MO, USA), prenylamine lactate (Sigma, St. Louis MO, USA).

Results

The influence of the calcium entry blockers on the audiogenic seizure response varied according to the different classes.

Intraperitoneal administration

Phenylalkylamines. No significant anticonvulsant effect was shown after verapamil (42–126 μmol/kg, i.p.) and D600 (8.4–21 μmol/kg, i.p.). In fact, some animals showed no tonic extension of hind feet. In contrast, a dose-dependent convulsant activity was observed shortly after administration of D 600 42 and 66 μmol/kg i.p., with rats showing clonus of the hind limbs, generalized myoclonus, vocalization, circling activity, barrel rolling and tonic flexion and more rarely extension of limbs. Respiratory arrest and death occurred before the application of auditory stimulus in 4/8 and 6/8 members of the groups treated with 42 and 66 μmol/kg of D 600 respectively.

Dihydropyridines. After i.p. administration of all dihydropyridine derivatives studied a dose-dependent reduction of the incidence of the audiogenic seizure phases was evident. In particular, 45 min following i.p. injection of nicardipine (105 μmol), nitrendipine (105 μmol), nisoldipine (105 μmol), nifedipine (126 μmol), nimodipine (126 μmol) and lacidipine (126 μmol) no tonic extension of limbs, nor tonus of neck, trunk and limb clonus (S.R.= 4–5) were observed. Sometimes a decreased incidence of clonic convulsions (S.R.= 2–3) was also observed. Significant protection (P<0.01) against the initial running phase was observed after i.p. administration of nitrendipine (126 μmol), nicardipine (126 μmol), nifedipine (126 μmol) and nimodipine (168 μmol) only. Reduction of locomotor activity, mild ataxia, piloerection and sedation was also evident in most of GEPRs treated with the highest doses of dihydropyridines. ED$_{50}$ values (± 95 confidence limits) for i.p. dihydropyridines against the tonic and clonic

phases of the audiogenic seizure response are reported in Table 1.

Dibenzothiazepine. Following diltiazem (84, 105, 126 and 168 μmol/kg i.p.), the tonic component of the audiogenic seizure (S.R.= 4, 5, 6, 7) decreased in a dose dependent manner in the absence of the behavioural or postural side effects. However, after lower doses (21, 42 and 66 μmol/kg i.p.) a decrease in the incidence of tonic extension of the hindfeet (S.R.= 8–9) was observed.

Diphenylalkylamines. Flunarizine (10.5–66 μmol/kg i.p.) and cinnarizine (66 and 168 μmol/kg i.p.), produced a dose-dependent reduction of the occurrence of the tonic phases (S.R.= 4,5,6,7,8,9) of the audiogenic seizure response 45 min after administration. Administration of 84 μmol/kg i.p. of flunarizine and 168 μmol i.p. of cinnarizine significantly reduced the incidence of the clonic phase (S.R.= 2–3), whereas a significant protection against the running phase was seen after flunarizine 105 and 126 μmol/kg and cinnarizine 210 μmol/kg only. Reduction of loco-

TABLE 1

ED$_{50}$ VALUES (WITH 95% CONFIDENCE LIMITS) OF SOME CALCIUM ANTAGONISTS ADMINISTERED I.P. AGAINST THE CLONIC AND TONIC PHASES OF AUDIOGENIC SEIZURES IN GEPRS

Drug	Clonic phase	Tonus of forelimbs
Cinnarizine	116(83–162)	64(49–84)
Flunarizine	69(40–118)	38(16–51)
Lacidipine	121(89–169)	50(34–74)
Nicardipine	95(68–133)	52(29–89)
Nifedipine	98(58–167)	56(31–99)
Nisoldipine	96(71–130)	42(38–64)
Nitrendipine	83(45–131)	42(26–67)
Nimodipine	92(66–129)	58(39–82)
HA 1004	136(109–170)	69(47–101)
Diltiazem	ND	78(63–96)
Prenylamine	ND	61(37–104)
Perhexiline	ND	99(62–158)

All data are expressed in μmol/kg. ND=not detectable.

motor activity, piloerection, sedation and ataxia was observed following the highest dose of fluna-rizine and cinnarizine. ED_{50} values ($\pm$ 95% confidence limits) for diphenylkylamines against the main phases of the audiogenic seizure response are reported in Table 1.

Prenylamine. Following prenylamine (66, 84, 105, 126 and 168 μmol/kg i.p.) the incidence of the tonic phases of the audiogenic seizures (S.R. = 4,5,6,7,8,8,9) was significantly modified in the absence of the behavioural or postural side effects. However, after lower doses (21 and 42 μmol/kg i.p.) only a decrease of incidence of hindfeet tonic extension (S.R. = 8–9) was observed. Some clonic jerks and hyperactivity was seen shortly after prenylamine (126–168 μmol/kg i.p.) administration. This symptomatology was followed 10–15 min later by ataxia with splayed hind limbs, reduced locomotor activity and pilo-erection. The ED_{50} value ($\pm$ 95% confidence limits) for prenylamine against sound-induced seizures is reported in Table 1.

Perhexiline. After perhexiline (42–168 μmol/kg i.p.) a dose-dependent decrease in incidence of the tonic component of audiogenic seizures was observed. No protection against the clonic phases (S.R. = 2–3) was observed after perhexiline 168–210 μmol/kg i.p. Reduction of locomotor activity, sedation, mild ataxia with splayed hind limbs and piloerection was also evident in animals treated with perhexiline 84, 126, 168 and 210 μmol/kg i.p. In addition, some clonic jerks and hyperactivity was seen shortly after perhexiline (168–210 μmol/kg i.p.) administration. This symptoma-tology was followed 10–15 min later by ataxia with splayed hind limbs, reduced locomotor ac-tivity and piloerection. The ED_{50} value ($\pm$ 95% confidence limits) for perhexiline against the tonic phase of sound-induced seizures is reported in Table 1.

HA 1004 (N-2-guanidinoethyl)-5-isoquinolinesul-fonamide). HA 1004 (42 and 66 μmol/kg i.p.) protected GEPRs against the phases 8 and 9 of audiogenic seizures. However, after HA 1004 84, 126 and 168 μmol/kg i.p. tonic (S.R. = 4,5,6,7) and clonic phases (S.R. = 2–3) of the seizure re-sponse were significantly depressed (P <0.05) al-though the presence of running phase (S.R. = 1) was observed. No ataxia or other side effects were evident after HA 1004 administration. The ED_{50} values ($\pm$ 95% confidence limits) for HA 1004 against the main phases of the audiogenic seizure are reported in Table 1.

Bay K 8644 (1,4-dihydro-2.6-dimethyl-3-nitro-4-(2-trifluorophenyl)pyridine-5-carboxylic acid.
Bay K 8644 (5.3–42 μmol/kg i.p.) did not signifi-cantly modify the occurrence of audiogenic seiz-ure phases. However, Bay K 8644 (5.3–42 μmol/kg i.p.) produced marked behavioural effects in GEPRs 3–8 min after the injection. Rats treated with 10.5, 21 and 42 μmol/kg i.p. showed a dose dependent ataxia with splayed hind limbs and delay of presence of the righting reflex (which was never completely lost). In addition, after a lat-ency period of 30–190 sec, depending on the dose, the rats showed arching back, tremor, ptosis, vo-calization, running, jumping, barrel rolling, scratching and clonus of fore limbs. These signs were followed after the highest doses (10.5, 21 and 42 μmol/kg i.p.) by tonic extension of both fore and hind limbs and generalized seizures fol-lowed by post ictal state. These phenomena were dose dependent and lasted 18–165 min. Clonic and tonic convulsions were observed in 4 out of 10, 8 out of 10, 10 out of 10 and 10 out of 10 rats treated with 5.3, 10.5, 21, and 42 μmol of Bay K 8644 respectively. Pretreatment (30 min before) with nimodipine or nitrendipine (84 μmol) was able to consistently reduce the incidence of the epileptic phenomena.

53

Intracerebroventricular administration

The anticonvulsant activity of some classes of calcium antagonists was also evaluated after i.c.v. administration of some hydrosoluble compounds (nicardipine, diltiazem, verapamil and HA 1004) (Table 2). Only nicardipine and HA 1004 showed a clear anticonvulsant activity.

Discussion

The present results show that some calcium antagonists given systemically or into the lateral cerebral ventricle possess anticonvulsant effects in a susceptible strain of rats (GEPR), thus confirming previous data in other models of epilepsy[4-6]. Recent findings show that in neuronal tissue, activation of dihydropyridine channels may modulate the release of neurotransmitter[13] and electrophysiological evidence has indicated that calcium ion currents in hippocampal pyramidal cells are sensitive to voltage-dependent calcium channel agonists and antagonists[14]. Moreover, central actions of voltage-dependent calcium channel blockers are indicated by their anticonvulsant activity in our model and by their effects on the withdrawal syndrome to ethanol and morphine in rodents[15,16].

The ED_{50} values and dose-response curves for various classes of calcium antagonists after i.p. administration revealed that flunarizine, a diphenylalkylamine calcium antagonist, was significantly more active than cinnarizine and dihydropyridine derivatives (Table 1). Phenylalkylamine derivatives (verapamil) belonging to Class I and believed to act similarly to diltiazem at the inner mouth of the Ca^{++} channel[8,17] were ineffective in preventing sound induced seizures in GEPRs (Tables 1 and 2).

Prenylamine, a compound belonging to class V, and perhexiline, belonging to class VI, possess anticonvulsant activity against the tonic phase only (Table 1). In addition, HA 1004, which inhibits Ca mobilization from intracellular stores and is described as a selective calmodulin antagonist[18], given i.p. was active as anticonvulsant (Tables 1 and 2), although its potency was 2.1–2.5

TABLE 2

THE EFFECT OF DILTIAZEM, HA 1004, NICARDIPINE AND VERAPAMIL ON AUDIOGENIC SEIZURES IN GERPS

Drug dose (μmol/kg)	Diltiazem	HA 1004	Nicardipine	Verapamil
	Median Seizure Score i.c.v. administration			
vehicle	9 (8)	9 (8)	9 (8)	9 (8)
0.05				9 (8)
0.1			6.3 (8)	9 (8)
0.5		8.1 (8)	4.8 (8)**	8.5 (8)
1.0	9 (8)	6.4 (8)	4.2 (8)**	4.9 (8)**
2.5			2.8 (8)***	
5.0	8.7 (8)	4.5 (8)**		
10.0	8.3 (8)	1.8 (8)***	1.3 (8)***	
25.0	8 (8)	0.5 (8)***	0.3 (8)***	

Groups of GEPRs were injected i.c.v. with the drugs or phosphate buffer (vehicle) and exposed to auditory stimulation 30 min after injection. Median seizure score for each dose level studied is reported. Number of rats is reported in brackets. Significant differences in the incidence of seizure phases between control (vehicle) and drug-treated groups are indicated by ** P <.01; *** P <.001 (Mann-Whitney U-test).

lower than that of flunarizine and almost similar (1.3–1.8 less potent) to that of some dihydropyridines. In addition, i.c.v. administration of nicardipine, a dihydropyridine derivative (0.5–25 μmol) and HA 1004 (5–25 μmol) was effective in preventing the clonic and tonic phases of audiogenic seizures (Table 2). The present findings show that there is a correlation between the systemic and the i.c.v. potency of some Ca^{++} antagonists against sound-induced seizures, as previously observed in DBA/2 mice[4]. This suggests that differences in anticonvulsant activity are more likely to be due to their different mechanism of action than to different pharmacokinetic properties, although a different distribution between the various drugs may also occur in different brain areas after i.c.v. administration according to their liposolubility properties. Flunarizine is in our study significantly more powerful than the dihydropyridines, suggesting that its pharmacokinetic properties make the compound more accessible to brainstem and mid brain structures involved in the pathophysiology of sound evoked seizures[19]. Much evidence points to a role of adenosine in the anticonvulsant activity of flunarizine: the drug inhibits the uptake of adenosine[20] and protection from audiogenic seizures in DBA/2 mice by flunarizine was reversed by aminophylline[3], an antagonist at adenosine receptors. In addition, Murphy and Snyder[21] reported that several 1,4-dihydropyridine derivatives (nifedipine, nitrendipine and nisoldipine) inhibited potently (IC_{50}'s < 10 μM) binding to adenosine receptors in the brain of the rat, whilst the non-dihydropyridine calcium channel blockers verapamil and diltiazem were very weak inhibitors of such binding (IC_{50}'s > 100 μM). The anticonvulsant activity of HA 1004, an intracellular calcium antagonist acting as a selective calmodulin antagonist suggests that an involvement of the Ca^{++}-calmodulin system may be important for the anticonvulsant effects of some calcium antagonists in the audiogenic seizures of GEPRs. This finding is quite interesting since the concentrations of phenytoin, carbamazepine and benzodiazepines which are effective against electroshock seizures correlate with their potency to inhibit the phosphorylation of some proteins regulated by a Ca^{++}-calmodulin system[22]. In addition, the present finding that Ca^{++} activation by Bay K 8644 induces in GEPRs epileptiform phenomena which are antagonized by nitrendipine and nimodipine indicates that Ca^{++} plays an important role in the pathophysiology of epileptic disorders and suggests that some Ca^{++} antagonists may be useful in the therapy of some types of human epilepsy.

Acknowledgements

Financial support from the Italian Ministry of Public Education (MPI, Rome) is gratefully acknowledged. We also thank Bayer S.p.A. Milan, Glaxo S.p.A., Verona, Janssen Farmaceutici, Roma and Sigma-Tau S.p.A., Pomezia (Rome), for their generous supply of calcium antagonists.

References

1 Heinemann, U., Lambert, J.D.C. (1986) In: Nisticò, G., Morselli, P.L., Lloyd, K.G., Fariello, R.G., Engel, J. Jr (eds.) Neurotransmitter, Seizures and Epilepsy, vol. III. Raven Press, New York, pp 65–73.
2 Pumain, R., Menini, C., Heinemann, U., Louvel, J., Silva-Barrat, C. (1985) Exp. Neurol. 89: 250.
3 De Sarro, G.B., Nisticò, G., Meldrum, B.S. (1986) Neuropharmacology 25: 695.
4 De Sarro, G.B., Meldrum, B.S., Nisticò, G. (1988) Br. J. Pharmacol. 93: 247.
5 Wauquier, A., Fransen, J., Clincke, G., Ashton, D., Edmonds, H.L. (1985) In: Godfraind, T., Vanhoutte, P.M., Govoni, S., Paoletti, R. (eds.) Calcium Entry Blockers and Tissue Protection. Raven Press, New York, U.S.A., pp. 163–172.
6 Overweg, J., Binnie, C.D., Meijer, J.W., Meinardi, H., Nuijten, S.T., Shmaltz, S., Wauquier, A. (1984) Epilepsia 25: 217.

7   Vanhoutte, P.M., Paoletti, R. (1987) Trends Pharmacol. Sci. 81: 4.

8   Glossman, H., Ferry, D.R., Goll, A., Striessing, J., Schober, M. (1985) J. Card. Pharmacol. 7: S20.

9   Murphy, K.M., Gould, R.J., Snyder, S.H. (1984) In: Scriabine, A., Vanon, S., Deck, K. (eds.). Nitrendipine. Urban and Schwarzenberg, Baltimore, U.S.A., pp. 107–117.

10 Jobe, P.C. (1981) In: Brown, R.D., Daigneault, E.A. (eds.) Pharmacology of Hearing. Experimental and Clinical bases. John Wiley, New York, U.S.A., pp 271–304.

11 Paxinos, G., Watson, C. (1986) The Rat Brain in Stereotaxic Coordinates. Academic Press, North Ryde, Australia.

12 Finney, D.J. (1978) Statistical Methods in Biological Assay, 3rd Edition, Ch. 18. Griffin, U.K.

13 Middlemiss, D.N., Spedding, M. (1985) Nature 314: 94.

14 Gahwiler, B.H., Brown, D.A. (1987) Neuroscience 20: 731.

15 Little, H.J., Dolin, S.J., Halsey, M.J. (1986) Life Sci. 39: 2059.

16 Baeyens, J.M., Esposito, E., Ossowska, G., Samanin, R. (1987) Eur. J. Pharmacol. 137: 9.

17 Godfraind, T. (1984) In: Godfraind, T., Hernier, A.G., Wellins, D. (eds.) Calcium Entry Blockers in Cardiovascular and Cerebral Dysfunctions. Martinus Nijhoff Publishers, Boston, U.S.A., pp 10–18.

18 Asano, M., Hidaka, H. (1985) J. Pharmacol. Exp. Ther. 234: 476.

19 Laird II, H.E., Jobe, P.C. (1987) In: Jobe, P.C., Laird, H.E. (eds.) Neurotransmitters and Epilepsy. Humana Press, Clifton, N.J., U.S.A., pp 57–89.

20 Phillis, J.W., Wu, P.H., Coffin, V.L. (1983) Gen. Pharmacol. 14: 475.

21 Murphy, K.M., Snyder, S.H. (1982) In: Merrill, G.F., Weiss, H.R. (eds.) Ca Entry Blockers, Adenosine and Neurohumors. Urban and Schwarzenberg, Baltimore, U.S.A., pp. 295–306.

22 De Lorenzo, R.J., Burdette, S., Holderness, J. (1981) Science 213: 546.

New Antiepileptic Drugs (Epilepsy Res. Suppl. 3)
Editors: F. Pisani, E. Perucca, G. Avanzini, A. Richens
© 1991 Elsevier Science Publishers B.V. (Biomedical Division)

CHAPTER 7

The brain distribution of old and new antiepileptic drugs

Francesco Monaco, Gianpietro Sechi

Neurological Clinic, Sassari University Medical School, Sassari, Italy

The specific site of action of antiepileptic drugs is the central nervous system (CNS), so that particular attention must be paid to the study of brain distribution of such substances. The main data for the major 'old' antiepileptic drugs and the few available for the new promising drugs will be illustrated in the following paragraphs.

Old drugs

1. Phenytoin

The concentration in brain is about 1 to 3 times the concentration of total drug in plasma and 6–10 times the free drug[1,2]. This is a result of binding to various subcellular fractions of brain cells[3]. More recent studies in animals indicate that a fraction of the protein-bound drug also penetrates the blood/brain barrier, possibly contributing to the amount of the cell-associated portion of the substance[4,5].

After intravenous (i.v.) injection in humans or intraperitoneal (i.p.) administration in rats, phenytoin rapidly enters the brain and reaches a peak level in less than 15 minutes, because of its high lipid solubility. Thereafter the concentration immediately falls as the plasma level declines as a result of redistribution phenomena, i.e. binding storage of the drug or deposit in other tissues (muscle, liver, fat or lungs). On continuous drug administration, these sites are saturated and the brain concentration again increases, paralleling the increase in plasma level.

Preferential accumulation of phenytoin in the superior and inferior colliculus, amygdala and hippocampus has been observed[6], but the functional significance of this difference is unknown.

In cats with convulsive generalized and focal penicillin-induced epileptic status a significant increase in the amount of phenytoin entering the brain was found, probably due to changes in cerebral blood flow, cerebral pH, vascular resistance, metabolic derangement and blood/brain barrier disruption[7]. On the other hand, Rapport et al.[8] observed that phenytoin concentration in the areas of maximum epileptogenic activity and astrogliosis in patients with epilepsy was much lower than the concentration in normal brain of controls, thus suggesting that reduced binding may results from astrogliosis, as glia concentrates phenytoin to a lesser extent than neurons.

The higher levels of phenytoin in white matter are in part a result of the high lipid content of this tissue. Phenytoin in fact accumulates in brain by avidly binding to brain proteins and phospholipids[9,10]. The binding to phospholipids depends on the partition coefficient (PC), which is high for phenytoin (log PC = 2.23) and is influenced by the Ca^{2+} concentration. The binding in brain may also be influenced by pH, since CO_2 administration increases the levels of phenytoin in the brain but lowers the levels of phenytoin in plasma[11].

Phenytoin distributes passively between plasma and cerebrospinal fluid (CSF). In brain interstitial fluid, the environment of neuronal and glial cells[12] in intimate and rapid communication with CSF[5,13], phenytoin concentration is low[14] and is not related to serum free concentration, thus demonstrating that the free phenytoin entering the brain accumulates into the cellular compartment.

2. Phenobarbitone

Lowering serum pH increases the non-ionized portion of the drug, thus enhancing diffusion into tissue, whereas higher serum pH has the opposite effect[15]. In brain, phenobarbitone binding correlates with the regional protein concentrations except for a slight additional affinity for sphyngomyelin[10]. In newborn cats, the highest concentration occurs in unmyelinated white matter, but by 8 weeks and in mature animals concentration is highest in gray matter[16]. In man, gray-to-white matter ratios vary from 0.86 to 1.11[17].

In cats rendered epileptic by parenteral penicillin and in cats with focal penicillin-induced epilepsy the brain penetration of phenobarbitone is impaired, although a gradual and progressive accumulation of the drug in brain tissue occurs[18]. In particular, phenobarbitone brain levels in cats with generalized penicillin epilepsy were lower than controls ($p < 0.5$); penetration was delayed, with brain binding occurring from time 30 minutes on. An even greater decrease and delay was observed in the focal epileptic brain ($p < 0.05$), the greater decrease being on the opposite side of the focus area.

3. Carbamazepine

In the rat, brain carbamazepine concentrations are 1.1 to 1.6 times those in the plasma, whereas the brain/plasma ratio for carbamazepine-10,11-epoxide ranges from 0.4 to 0.8, suggesting reduced brain penetration[19]. Studies on the regional brain distribution show carbamazepine concentrations in the cortex, thalamus and hippocampus shortly after dosing, while later subcortical structures have the highest relative concentrations, and such distribution is similar to that observed for other lipophilic drugs[20]. Carbamazepine-10,11-epoxide appears to have a uniform distribution[21] too.

Carbamazepine brain uptake in cats with generalized penicillin epilepsy was decreased and delayed ($0.005 < p < 0.025$)[22], whereas the distribution of the drug into the penicillin epileptic focus did not differ significantly from the other areas of the brain[23].

4. Valproic acid

Valproic acid enters the brain and CSF rapidly, with brain peak levels reached within minutes[24], with a subsequent decline of drug concentration paralleling that in plasma, indicating a ready equilibration between brain and capillary blood.

The only available data on valproic acid brain concentrations are from the biopsy study of Vajda et al.[25], who found relatively low levels (6.8–27.8% of total plasma levels). Thus, unlike many of the aromatic or heterocyclic antiepileptic drugs such as phenytoin and phenobarbitone, which typically exhibit brain/CSF concentration ratios well exceeding unity, valproic acid does not appear to concentrate in brain tissue[26]. Goldberg and Todoroff[27] have in fact reported that valproic acid does not exhibit binding to rabbit brain homogenates or subcellular fractions of brain tissue.

The transport of valproic acid across the blood/brain barrier and blood-CSF barrier is carrier-mediated[5], as existing evidence suggests that valproic acid utilizes the same active transport system for the removal of acidic metabolites of monoamines from CNS[28]. Valproic acid is also transported out of CSF across the choroid plexus.

Data on brain concentrations or CNS kinetics of the mono-unsatured metabolite Δ^2-valproic acid are controversial. Slow accumulation of this metabolite in substantia nigra, superior and inferior colliculus, hippocampus and medulla of rat brain may provide an explanation for the slow reversibility of the anticonvulsant effect of valproic acid[4].

5. Benzodiazepines

Benzodiazepines cross the blood/brain barrier easily and distribute rapidly in the brain[20]. As a result of its low pKa and its lipophilic character, diazepam distributes quickly in lipid tissues and brain[30]. A subsequent decline and selective binding in different brain regions and redistribution may occur. Similar data are obtained for clonazepam[31].

New drugs

1. Vigabatrin

Vigabatrin is an irreversible suicide inhibitor of GABA-transaminase (GABA-T) and a potent anticonvulsant both in animal models of epilepsy and in humans. The S-enantiomer of vigabatrin is taken up by a high affinity carrier ($K_m = 80$ μM) into neurons but not in astrocytes, which have only a low affinity ($K_m > 500$ μM) transport carrier. It is likely that the compound is transported by a carrier different from the GABA carrier[32]. The preferential uptake and inhibitory effect of vigabatrin in neurons is in keeping with the observation that prolonged treatment with the drug in vivo leads to a preferential increase in GABA levels in the nerve endings compared to whole brain[33,34].

The drug enters the brain rapidly, with maximum levels at 30 minutes after an i.p. dose of 1500 mg/kg in mice[35]. Due to irreversible binding to GABA-T, vigabatrin may persist in the brain for a long time, a phenomenon which is probably responsible for its long-lasting effect[36]. The half-life of vigabatrin in the brain is about 16 hours and the drug is still detectable 3 days after injection.

2. Gabapentin

Absorption, metabolism and excretion of gabapentin were investigated in rats, dogs and healthy volunteers following oral administration of ^{14}C-labelled substance[32]. Following i.v. administration to rats, similar blood and brain concentrations were observed after a short distribution phase (Table 1).

3. Stiripentol

Radioactivity was measured in CNS of Wistar rats following i.v. administration of ^{3}H-stiripentol[38]. The maximum activity was noted in the cerebellum.

TABLE 1

GABAPENTIN CNS CONCENTRATIONS FOLLOWING I.V. ADMINISTRATION OF 25 MG/KG ^{14}C-LABELLED GABAPENTIN TO RATS

Brain region	Concentration (μg/g), mean ± SD			
	15 min	1 hr	2 hr	4 hr
Cerebrum	8.55±0.96	11.78±2.21	4.99±0.88	2.06±0.53
Cerebellum	11.02±2.24	11.92±0.8	5.18±0.78	1.92±0.49
Medulla oblongata	8.79±1.04	10.53±0.67	4.86±0.9	1.84±0.75

(N=5)
Data from Vollmer et al.[37].

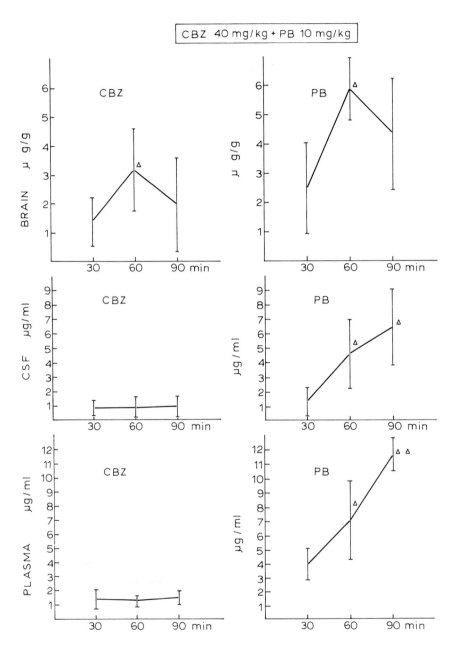

Fig. 1. Brain, CSF and plasma concentrations (means ± SD) of carbamazepine (CBZ; 40 mg/kg) and phenobarbitone (PB; 10 mg/kg) after intraperitoneal administration to normal cats. One triangle, $p < 0.05$; two triangles, $p < 0.01$, compared with value at 30 min. (Reproduced from Monaco et al.[44]).

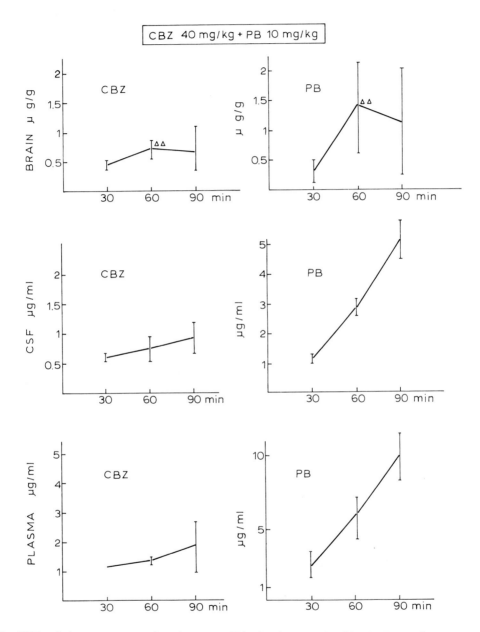

Fig. 2. Brain, CSF and plasma concentrations (means ± SD) of carbamazepine (CBZ; 40 mg/kg) and phenobarbitone (PB; 10 mg/kg) after intraperitoneal administration of the drugs in combination to cats rendered epileptic by parenteral penicillin. One triangle, p < 0.05; two triangles, p < 0.01, compared with value at 30 min. (Reproduced from Monaco et al.[44]).

CBZ

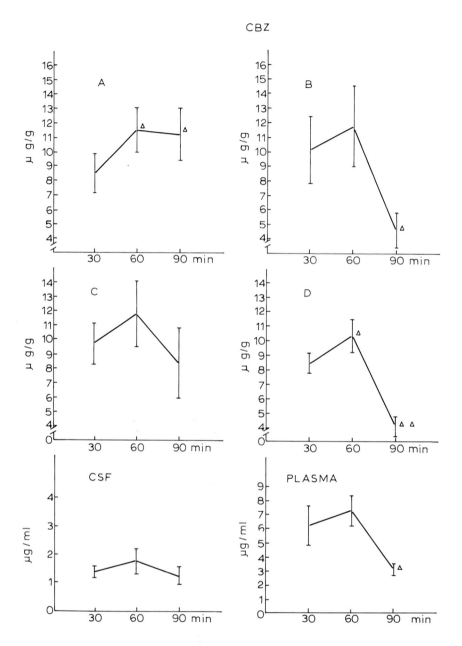

Fig. 3. Brain, CSF and plasma concentrations (means ± SD) of carbamazepine after intraperitoneal administration of the drug in combination with phenobarbitone to cats with focal penicillin-induced epilepsy. Brain regions: A) the focus area, B) area contralateral to the focus, C) homolateral posterior area to the focus, and D) contralateral posterior area to the focus. One triangle, $p < 0.05$; two triangles, $p < 0.01$, compared with value at 30 min. (Reproduced from Monaco et al.[44]).

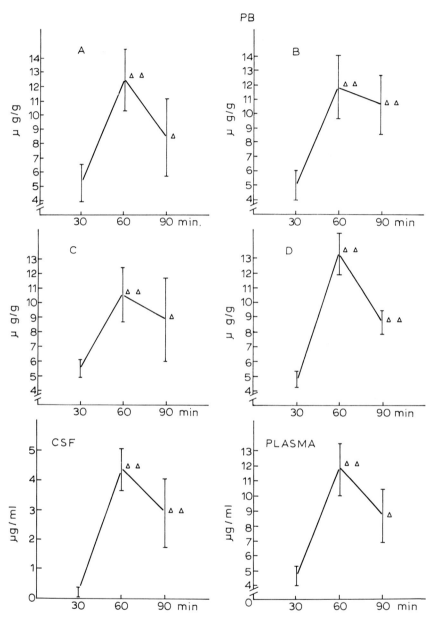

PB

Fig. 4. Brain, CSF and plasma concentrations of phenobarbitone after intraperitoneal administration of the drug in combination with carbamazepine to cats with focal penicillin-induced epilepsy. Brain regions: A) the focus area, B) area contralateral to the focus, C) homolateral posterior area to the focus, and D) contralateral posterior area to the focus. One triangle, $p < 0.05$; two triangles, $p < 0.01$, compared with value at 30 min. (Reproduced from Monaco et al.[44]).

Conclusions

Löscher and Frey[39] showed in anesthetised dogs that in physiological conditions the permeability of the blood-CSF and hence the blood/brain barrier is similar if not identical for all drugs, and the modality of penetration is governed by three physico-chemical properties: a) the degree of ionization; b) the liposolubility, and c) the plasma protein binding. The most important property is the liposolubility. Only valproic acid behaves differently, similar to other weak acids (such as acetylsalicylic acid and penicillin) for which active transport has been shown in various species[5].

During severe convulsions changes in cerebral blood flow, brain pH, vascular resistance, blood/brain barrier disruption and metabolic derangement may affect the distribution of drugs[40-43]. To date, we have no evidence to indicate which of these factors plays the most important role, and presumably all of them interact in altering the brain drug uptake. In our opinion, epileptic seizures may lead to increased penetration of drugs into the cerebral compartment through the disrupted blood/brain barrier, but the distribution and binding of different drugs vary according to their ability to prevent the metabolic alterations associated with severe convulsions.

When phenobarbitone is given in combination with carbamazepine, the brain concentrations of the drug are always lower than the control levels in all experimental models of epilepsy but the focal model[44] (Figures 1–4). We speculated that carbamazepine somehow facilitated entry into the brain of phenobarbitone by preventing the binding impairment. That event, however, did not potentiate the anticonvulsant activity of carbamazepine. Thus, from a pharmacodynamic point of view, there is no clear experimental evidence that a combination of two drugs is better than one drug alone in controlling epilepsy.

References

1 Firemark, J., Barlow, C.F., Roth, J. (1976) Int. J. Neuropharmacol. 2: 25–38.
2 Friel, P.N., Ojemann, G.A., Rapport, R.L., Levy, R.H., Van Belle, G. (1989) Epilepsy Res. 3: 82–85.
3 Woodbury, D.M. (1980) Adv. Neurol. 27: 447–471.
4 Löscher, W., Nau, H.H. (1983) J. Pharmacol. Exp. Ther. 226: 845–854.
5 Cornford, E.M., Oldendorf, V.H. (1986) In: Delgado-Escueta, A.V., Ward, A.A. Jr, Woodbury, D.M., Porter, R.J. (eds), Basic Mechanisms of the Epilepsies: Molecular and Cellular Approaches. Raven Press, New York, pp 787–812.
6 Nakamura, K., Masuda, Y., Nakatsuji, K., Kiroka, T. (1966) Naunyn Schmiedebergs Arch. Pharmacol. 254: 406–417.
7 Sechi, G.P., Russo, A., Rosati, G., Mutani, R., Monaco, F. (1987) Epilepsy Res. 1: 173–177.
8 Rapport, R.L., Harris, B., Friel, P.N., Ojemann, G.A. (1975) Arch. Neurol. 32: 549–554.
9 Goldberg, M.A., Crandall, P.H. (1978) Neurology 28: 881–885.
10 Goldberg, M.A. (1980) Adv. Neurol. 27: 323–337.
11 Woodbury, D.M. (1969) Epilepsia 10: 121–124.
12 Felghenhauer, K. (1986) J. Neurol. 233: 193–194.
13 Partridge, W.M. (1986) Ann. Intern. Med. 105: 82–95.
14 Sechi, G.P., Petruzzi, V., Rosati, G., Tanca, S., Monaco, F., Formato, M., Rubattu, L., De Riu, P. (1989) Epilepsia 30: 235–239.
15 Wade, A. (1980) In: Pharmaceutical Handbook, The Pharmaceutical Press, London.
16 Roth, L.J., Barlow, C.F. (1961) Science 132: 22–31.
17 Harvey, C.D., Sherwin, A.L., Van Der Kleijn, E. (1977) Can. J. Neurol. Sci. 4: 89–92.
18 Monaco, F., Sechi, G.P., Piredda, S., Frassetto, M., Zoroddu, G.F., Russo, A., Mutani, R. (1983) Epilepsia 24: 686–691.
19 Morselli, P.L., Gerna, M., Garattini, S. (1971) Biochem. Pharmacol. 20: 2043–2047.
20 Placidi, G.F., Tognoni, G., Pacifici, G.M., Cassano, G.B., Morselli, P.L. (1976) Psychopharmacologia 48: 133–137.
21 Pantarotto, C., Crunelli, V., Lanzoni, J., Frigerio, A., Quattrone, A. (1979) Ann. Biochem. 93: 115–123.
22 Monaco, F., Piredda, S., Traccis, S., Frassetto, M., Mutani, R. (1982) Ital. J. Neurol. Sci. 2: 107–110.
23 Monaco, F., Piredda, S., Sechi, G.P., Frassetto, M., Mutani, R. (1982) Epilepsia 23: 615–618.
24 Hariton, C., Ciesielski, L., Simler, S., Valli, M., Jadot,

C., Gobaille, S., Mesdjian, E., Mandel, P. (1984) Biopharm. Drug. Disp. 5: 409–414.

25 Vajda, F.J.E., Donnan, G.A., Phillips, J., Bladin, P.F. (1981) Neurology 31: 486–487.

26 Levy, R.H., Shen, D.D. (1989) In: Levy, R.H., Dreifuss, F.E., Mattson, R.H., Meldrum, B.S., Penry, J.K. (eds.) Antiepileptic Drugs, Third Edition. Raven Press, New York, pp. 583–599.

27 Goldberg, M.A., Todoroff, T. (1980) Neurology 30: 826–831

28 MacMillan, V. (1979) Can. J. Physiol. Pharmacol. 57: 843–847.

29 Garattini, S., Marcucci, F., Morselli, P.L., Mussini, E. (1973) In: Davies, D.S., Prichard, B.N.C. (eds.) Biological Effects of Drugs in Relation to their Plasma Concentrations. Macmillan, London, pp 211–226.

30 Ramsay, R.E., Hammond, E.J., Perchalski, R.J., Wilder, B.J. (1979) Arch. Neurol. 36: 535–539.

31 Sato, S. (1989) In: Levy, R.H., Dreifuss, F.E., Mattson, R.H., Meldrum, B.S., Penry, J.K. (eds.) Antiepileptic Drugs, Third Edition. Raven Press, New York, pp. 765–784.

32 Larsson, O.M., Gram, L., Seiler, N., Schousboe, A. (1985) Acta Neurol. Scand. 66: 72.

33 Gale, K. (1984) Life Sci 34: 701–706.

34 Mangione, M.C., Ferraro, T.N., Fariello, R.G., Garant, D.S., Golden, G.T., Hare, T.A. (1987) Epilepsia 28: 583.

35 Jung, M.J., Lippert, B., Metcalf, B.W., Bolen, P., Schechter, P.J. (1977) J. Neurochem. 29: 797–802.

36 Jung, M.J. (1978) In: Seiler, N., Jung, M.J., Koch-Weser, J. (eds.) Enzyme Activated Irreversible Inhibitors. Elsevier, Amsterdam, pp 135–148.

37 Vollmer, K.O., Von Hodenberg, A., Kolle, E.U. (1986) Azneimittel. Forsch. 36: 830–839.

38 Pieri, F., Wegmann, R., Astoin, J. (1982) Eur. J. Drug Metab. Pharmacokinet. 7: 5–10.

39 Löscher, W., Frey, H.H. (1984) Epilepsia 25: 346–352.

40 Bolwig, T.G., Hertz, M.M., Paulson, O.B., Spotoft, H., Rafaelsen, O.J. (1977) Eur. J. Clin. Invest. 7: 87–93.

41 Doba, N., Beresdorf, H.R., Reis, D. (1975) Brain Res. 90: 115–132.

42 Duffy, T.E., Howse, D.C., Plum, F. (1975) J. Neurochem. 24: 925–934.

43 Petito, C.K., Schaefer, J.A., Plum, F. (1977) Brain Res. 127: 251–267.

44 Monaco, F., Sechi, G.P., Russo, A., Traccis, S., Mutani, R. (1985) Epilepsia 26: 103–108.

Clinical Aspects

New Antiepileptic Drugs (Epilepsy Res. Suppl. 3)
Editors: F. Pisani, E. Perucca, G. Avanzini, A. Richens
© 1991 Elsevier Science Publishers B.V. (Biomedical Division)

CHAPTER 8

Evaluation of clinical efficacy in antiepileptic drug trials

Dieter Schmidt

Department of Neurology, Free University of Berlin, Germany

Introduction

The clinical evaluation of new antiepileptic drugs deserves special interest for several reasons. Quite a number of new antiepileptic drugs are currently undergoing early phases of clinical testing. Advances in the pre-clinical testing of antiepileptic drugs will increase the number of future components for clinical testing[1]. New guidelines for clinical testing of antiepileptic drugs have recently been presented providing a framework of suggestions for flexible and adaptive evaluation of new antiepileptic drugs[2]. The guidelines suggest answers to questions about the efficacy of new compounds in a stepwise logical fashion. In this chapter various issues of the evaluation of clinical efficacy in antiepileptic drug trials will be discussed: seizure control, the use of placebo and active drug as control, endpoints of efficacy, patients with incomplete control of seizures, previously untreated patients, single drug designs, and finally add-on drug study designs.

Seizure control

Seizure control is a pivotal measure of efficacy for a new antiepileptic drug. Most studies have used seizure counts as the main outcome variable. These can be analysed as number of seizures over a time interval, interval between seizures, changes in seizure pattern, and duration of seizures (Table 1).

A number of difficulties can arise with the employment of these parameters. The number of seizures and their duration, and the interval between seizures over a time interval can be most reliably established during continuous 24 hour Video-EEG-monitoring. This approach has been used for measurement of efficacy in patients with infantile spasms, myoclonic seizures, absence seizures and various seizures types of patients with Lennox-Gastaut Syndrome. In-patient observation and a high number of seizures during the observation period of several days are required. Even with sophisticated Video-EEG-monitoring it may be difficult to exactly define the duration of a seizure, especially during complex partial seizures which may gradually blend into the pre-trial behaviour. Individual patients may

TABLE 1

EVALUATION PARAMETERS OF CLINICAL EFFICACY

Change in seizure frequency
Change in seizure-free interval
Change in seizure pattern
Change in functional capacity
Percentage of responders
Treatment failures

70

have complex partial seizures lasting for 10–15 minutes.

For tonic-clonic seizures the end of the seizure can be detected more easily. The postictal state is heralded by cessation of clonic activity and disappearance of the movement artefact on the EEG. Variability of the duration of tonic-clonic seizures is smaller than in complex partial seizures. Most tonic-clonic seizures last 1–2 minutes. Complex partial seizures offer another problem, i.e. they may occur in clusters. In these clusters, it may be difficult to exactly define the postictal state and the onset of the next seizure. The exact number of complex partial seizures during a cluster may be difficult to determine. If ten complex partial seizures occur during four hours, is it appropriate to count this episode as ten seizures or one seizure? The issue of clustering needs to be addressed stating how the investigator handled episodes of clustering seizures.

Another difficult issue is the counting of simple partial seizures. Simple partial seizures may not be observable. One needs to rely on the reports of the individual with seizures. In addition simple partial seizures may occur very frequently up to 30–50 per day and may outrank the number of complex partial seizures or generalized tonic-clonic seizures. It is therefore preferable to count the number of seizures separately for each type of seizure. Antiepileptic drugs may reduce the number of auras without reducing the number of complex partial seizures or vice versa.

The recognition that one type of seizure may occur in several different epilepsies or epileptic syndromes has added to the complexity of anti-epileptic drug trials. There is a clear need to learn more about the drug response of the same type of seizure occurring in various epilepsy syndromes. One example are generalized tonic-clonic seizures. They may occur in idiopathic generalized epilepsies, in cryptogenic and symptomatic generalized epilepsies and in epilepsies which cannot

be established as focal or generalized. This latter group includes patients with generalized tonic-clonic seizures during sleep or patients with photosensitivity and a focal interictal paroxysmal EEG activity. Valproate is the standard drug for patients with idiopathic generalized epilepsies while phenytoin is not recommended. Yet both drugs are active against generalized tonic-clonic seizures in cryptogenic generalized epilepsies or focal epilepsies. It is therefore useful to establish not only the type of seizure according to the International classification of seizures[3] but also, if possible, to outline the epilepsy or the epileptic syndrome according to the International Classification of Epilepsies and Epileptic Syndromes[4].

Placebo and active drug as controls

A placebo or an active drug is commonly chosen as a control to compare the clinical efficacy of a novel antiepileptic drug. Even in patients with chronic intractable epilepsy treated for several years, placebo has considerable efficacy to reduce the number of seizures. In a recent double blind, parallel group study addition of placebo lead to a 50% reduction of partial seizure frequency in 9.8% of the patients. The median reduction in partial seizure frequency during the 12 week trial was 12.5% for placebo and 29.2% for the addition of the novel antiepileptic drug[5]. While the use of a placebo control provides valuable data for comparison of efficacy with the novel antiepileptic drug, it may pose an ethical dilemma in drug trials with previously untreated patients with tonic-clonic seizures. It has been estimated that about 70% of all patients will develop further seizures, i.e. epilepsy following a single generalized tonic-clonic seizure[6]. Introduction of placebo would expose these patients to further tonic-clonic seizures. Status epilepticus would be another example where placebo controls seem to carry an untoward risk. In contrast, previously untreated patients with absence seizures or complex partial seizures may benefit from placebo controls given the unknown natural history of untreated individuals and the relatively low risk of further seizures associated with placebo control during a trial of several weeks to months.

An active control drug offers valid information about the comparison of the novel drug with the standard drug yet active control investigations harbour a hazard. If there is no difference in efficacy between active control and the novel drug the proportion of patients who actually benefit from either drug is not known[7]. A recent example is a double-blind single drug trial comparing oxcarbazepine, a novel drug, with carbamazepine as a standard drug for treatment of newly diagnosed epilepsy with partial seizures[8]. There was no difference between the two drugs. The most plausible explanation for this finding is a pharmacological effect of the experimental drug, but this argument needs support from additional external information. In the case of oxcarbazepine, placebo controlled trials are not yet available. While this argument may sound a bit far-fetched, it does point to a lack of knowledge regarding the proportion of patients with active epilepsy who benefit from taking a standard drug e.g. carbamazepine. Future studies will need to address this question. In patients with intractable epilepsy active control investigations are fraught with the same uncertainty of how many patients do actually benefit from taking a standard drug. Again, this question needs to be addressed in future trials.

Endpoints of efficacy

A reduction of 50% is generally accepted as evidence for clinical efficacy of an antiepileptic drug, even though most patients would agree that such a reduction, while worthwhile, is not fully satisfactory. However, even standard drugs when used as additional treatment will achieve a 50%

TABLE 2

SINGLE DRUG THERAPY IN PARTIAL EPILEPSY: COMPLETE SEIZURE CONTROL

Previously untreated patients[18]	
68 %	Carbamazepine or
66 %	Phenytoin
Previously untreated and undertreated patients[10]	
65 %	Carbamazepine or
33 %	Phenobarbitone or
34 %	Phenytoin or
26 %	Primidone
42 %	All
Previously undertreated patients[21]	
31 %	Phenytoin or
	Primidone
Previously intractable patients[22]	
12 %	Phenytoin or
	Carbamazepine or
	Phenobarbitone

reduction in no more than one-third of patients with intractable epilepsy (Table 2). These data provide a useful basis for comparison with novel antiepileptic drugs. Furthermore, they point out the limited value of two drug treatment in failures of adequate single drug therapy[6].

A very convincing endpoint of clinical efficacy in studies with previously untreated patients is remission of seizures. Usually a 6 month, or 1 year remission is established in longterm drug trials. In a recent comparative trial in 241 previously untreated epileptic patients 82% of all patients had achieved a 6 month remission and 60% a 1 year remission. The median duration of follow up was 26 months[9]. Another endpoint, no less valuable is treatment failure e.g. persistence of seizures or seizures plus toxicity. Among 421 previously untreated or undertreated epileptic patients seizures alone were reasons for drug failure in 11 patients only, while toxicity alone or toxicity plus seizures were responsible for treatment failure in 127 and 85 patients, respectively[10]. The most significant prognostic feature was the number of pretreatment tonic-clonic or partial seizures[9].

Such large-scale studies are very valuable, and usually are performed at the end of the clinical development of a compound or even during phase IV. There is a clear need however to estimate the efficacy of an antiepileptic agent very early, preferably at a time when single dose safety is available.

Interictal spike counting

In the search for pilot efficacy data suppression of interictal spikes has been proposed as a method for assessing the antiepileptic effect of drugs. The EEG spike is the hallmark of interictal epileptic activity. The relationship between spikes and seizures is uncertain however, and for many patients the correlation is poor. An increase of interictal spiking following a seizure or during drowsiness is well recognized. Milligan and Richens[11] pointed out further shortcomings: the difficulty in counting large numbers of spikes; drugs with slow onset of action will require prolonged periods of counting: patients with frequent spikes are not common. A major difficulty is the considerable intra-individual and inter-individual variation in the number of spikes. The degree of variation is such that minor drug effects are obscured by the background 'noise'. Spike activity is profoundly increased if the drug under test may precipitate drowsiness. On the positive side, single doses of benzodiazepines, phenytoin, lamotrigine[12] have suppressed spikes. In spite of these limitations, spike counting may provide an early indication that the compound is a potential antiepileptic drug, although a negative response may not be taken a strong indicator of lack of antiepileptic efficacy[13].

Patients with incomplete control of seizures

The Commission on Antiepileptic Drugs[4] put forward a definition of incomplete seizure control as seizures despite maximum tolerable serum concentrations of standard drug therapy. Good drug compliance is a prerequisite. The National Association of Epilepsy Centers has defined as intractable a person whose seizures do not come under control after nine months of treatment under the care of a neurologist. Another proposal includes the persistence of an unacceptable quantity of epileptic seizures despite medical management. The advantage of the Commission's proposal is its operational criteria. 'Maximum tolerable serum concentrations' is clinically defined through the development of drug toxicity during treatment precluding a further dose increment. Standard drug therapy reflects the choice of carbamazepine, phenytoin, phenobarbitone, primidone or valproic acid for partial seizures or secondarily generalized tonic-clonic seizures while absence seizures are treated with valproic acid and ethosuximide. In patients with idiopathic generalized epilepsy, generalized tonic-clonic seizures are treated with valproic acid. Reliable measurement of drug compliance may be rather difficult[14]. Poor drug compliance may be detected by a high variation of repeated serum concentration values or with the help of a drug dispenser which can be monitored electronically. The question 'Did you take your drug as prescribed and how often did you miss your drug since the last visit' asked by the physician or the clinical nurse has not been replaced as yet however.

The assessment of the response to previous treatment is required by the guidelines for clinical testing of antiepileptic drugs. An antiepileptic drug can be expected to be more effective in an untreated or undertreated patient compared to a patient unresponsive to maximum tolerable sin-

TABLE 3

MEDICAL INTRACTABILITY SCALE: THE SCALE ALLOWS FOR ASSESSMENT OF THE RESPONSE TO PREVIOUS
DRUG TREATMENT FOR EACH TYPE OF SEIZURE IN THE INDIVIDUAL PATIENT[6]

Index of Intractability	Seizures persist despite treatment with
0	Other than primary drug regardless of its daily dose
1	Primary drug below the recommended daily dose
2	Primary drug within the recommended daily dose
3	Primary drug with plasma concentrations within the recommended 'therapeutic range'
4	Primary drug with maximum tolerable daily dose
5	More than one primary drug with maximum tolerable daily dose in subsequent single-drug therapy for at least nine months

gle and multiple drug therapy for one or more years[6] (Table 2). Recently, a scale was proposed for the quantification of the response to previous drugs (Table 3).

Previously untreated patients

Previously untreated patients with epilepsy clearly benefit most from standard drug therapy and may be expected to benefit more than intractable cases (Table 2). A potentially high spontaneous remission rate, difficult acquisition of patients, preliminary diagnostic classification and the ethical issues are some of the problems encountered in drug trials of previously untreated patients. The current guidelines offer a number of instances where previously untreated patients may be exposed to investigational drugs (Table 4).

Single drug designs

Single drug treatment offers several advantages and some disadvantages (Table 5). The effect of single drug therapy in partial epilepsy clearly shows that standard drugs perform better in single drug trials than in combinations treatment patients with intractable epilepsy (Table 6). For ad-

ditional discussion see Schmidt[6].

From the above results it is clearly useful to test a new component as early as possible in single drug trials. The current guidelines offer this possibility (Table 7).

One intriguing aspect of single drug therapy in intractable epilepsy is our lack of precise knowledge on what proportion of patients with incomplete control of seizures actually benefit from current antiepileptic drug treatment. This question has surprisingly enough never been studied systematically. Anecdotal evidence of individual patients with intractable partial seizures suggests that some patients do just as well or even better without antiepileptic drug[6]. Another one may be that withdrawal of drugs during presurgical evaluation does not invariably lead to an increase of seizures. Furthermore poor compliance leads to exacerbation of seizures in only about one third to one half of all patients[15]. Finally, long term follow up of patients with epilepsy suggests that about 75% of patients do enter into remission, but only half of these patients were taking antiepileptic drugs, while 52% were on drug treatment. A surprising observation was that antiepileptic drug therapy was a variable not necessarily independently and causally associated with seizure outcome in the long run[16]. It is therefore very

TABLE 4

SELECTION OF PREVIOUSLY UNTREATED PATIENTS FOR ANTIEPILEPTIC DRUG TRIALS

1	Previously untreated patients can be randomized to receive the new drug as compared with placebo or a reference drug. If placebo control is used, analysis of seizure-free interval is recommended during late phase I.
2	If clear proof of efficacy has emerged for a drug with severe drug interactions, previously untreated patients can be evaluated in a monotherapy design. The interval to next seizure should be the variable evaluated. An add-on design for previously untreated patients is a blinded parallel-group design in which an established antiepileptic drug is administered systematically in association with either the new drug or a placebo according to a randomized sequence during late phase II.
3	Previously untreated patients should be included in phase III trials if the new drug has shown evidence of efficacy. Variable randomized designs may be used, involving open-label or blind procedures, comparing the new drug with placebo or a reference drug during phase III.
4	Previously untreated patients may be included in accordance with the appriated medication during phase IV.

TABLE 5

PRESUMED ADVANTAGES AND DISADVANTAGES OF SINGLE DRUG THERAPY

Advantages	Disadvantages
– efficacious in about 50–80 % of patients with focal epilepsy	– not effective in about 20–50 %
– no drug interactions	– we do not know how many patients with uncontrolled focal
– better compliance	epilepsy actually benefit from any drug treatment
– analysis of drug action is easier	– specificity of primary drugs not known, if any

TABLE 6

RESPONSE TO ADDITION OF STANDARD ANTIEPILEPTIC DRUGS IN PATIENTS WITH INTRACTABLE EPILEPSY. EXPRESSED AS PERCENT OF PATIENTS WITH 50 % REDUCTION IN SEIZURE FREQUENCY

Additional drug	Percent of patients with 50 % reduction
Phenytoin, Carbamazepine or Phenobarbitone[20]	17 %
Carbamazepine[23]	27 %
Phenytoin[23]	11 %
Valproate[23]	30 %
Valproate[24]	37 %

TABLE 7

SINGLE ANTIEPILEPTIC DRUG TRIALS

1	If pronounced drug interactions occurred in late phase I studies, monotherapy trials may be performed during early phase II.
2	If clear proof of efficacy has emerged for a drug with severe drug interactions a monotherapy design can be used during late phase II.
3	During phase III particular attention should be paid to monotherapy designs.

75

difficult to predict the response to a new drug (or a conventional drug) in patients with incomplete control of seizures. When patients with incomplete control of seizures are randomized in a double-blind fashion to continue on their previous single drug or to receive placebo, one would find out what proportion in fact requires further drug treatment.

Two drug designs

The notion that two drugs are better than one has been proposed as early as 1885 by Gowers[17] and is today still traditionally accepted by many (Table 8). This is surprising because there is no persuasive experimental evidence in animals that two drugs can achieve an antiepileptic effect that cannot be produced by either drug alone[6] nor is there any sound support to favour any specific combinations of two drugs. In a recent trial 60% of the patients were adequately managed by monotherapy with the first drug (carbamazepine, phenytoin, phenobarbitone or primidone) they received by random assignment. Treatment failures with the first drug were switched to a second

drug (either because of toxicity, lack of seizure control, or both) and 55% were successfully treated with the second randomly assigned drug also used as monotherapy. In this study, 20% could not be adequately managed on monotherapy. Of this 20%, 49% could be managed by treatment with two drugs in combination[10].

The clinical evidence is marred by the fact that no controlled trials are available (Table 2). Two drug therapy with standard drugs of first choice will result in moderate reduction of seizure frequency with complete control in about 11–20 % of the patients[18,19,20]. How does the reduction in seizures on addition of standard drugs compare with that of addition of novel antiepileptic drugs? Novel antiepileptic drugs can be assessed by add-on designs in patients with intractable epilepsy (Table 9). Furthermore, differences between individual new antiepileptic agents can be detected. Clobazam, while very effective initially will lose its efficacy in as many as two thirds of the patients, but still would be effective in about 23%. Vigabatrin has been shown to reduce seizure by 50% in 45% of patients with intractable epilepsy. Gabapentin and lamotrigine offer still worth-

TABLE 8

PRESUMED ADVANTAGES AND DISADVANTAGES OF TWO DRUG THERAPY

Advantages	Disadvantages
– efficacious in non-responders to single drug therapy: proportion unknown, may be as high as 30–50 %	– drug interactions leading to reduced efficacy and increased toxicity
	– poorer compliance
– eases off pressure 'to do something' in uncontrolled epilepsy	– no sound experimental evidence, equivocal clinical evidence
– is traditionally accepted	
	– confounds analysis of single drug action
	– more expensive

TABLE 9

TREATMENT RESPONSE TO NOVEL ANTIEPILEPTIC DRUGS

Additional drug	Percent of patients with 50 % reduction
Clobazam[25]	70 %
Vigabatrin[26]	45 %
Gabapentin[5]	25 %
Lamotrigine[27,28]	25 %
	10 %
Progabide[29]	5 %
Placebo[5]	12.5 %

while, but less impressive results, while progabide clearly is least effective among the new antiepileptic drugs shown here. Finally, it is useful to know that a placebo response in 12.5% can be expected.

Future perspectives

A fruitful discussion of the merits and disadvantages of two drug therapy has been hampered by the lack of concepts regarding rational two drug therapy. A number of strategies for rational two drug therapy need to be assessed in future trials (Table 10).

Finally, the current concept of clinical efficacy covers mainly the development of new drugs which suppress seizure manifestations in patients with active epilepsy. In addition, we need to de-

TABLE 10

Strategies for rational two drug therapy

Combine drugs which differ in their mechanism of action
Combine drugs which show mutually beneficial pharmacokinetics
Combine drugs which differ in their specific effect on various seizure types to improve efficacy
Combine drugs with different side effects to avoid additive or supraadditive toxicity

velop new drugs which limit or abolish the process of epileptogenesis, i.e. prevent epilepsy following a single seizure or ideally even before the first seizure occurs. In that sense clinical testing of novel antiepileptic drug trials provides challenges for the future.

References

1 Löscher, W., Schmidt, D. (1988) Epilepsy Res. 2: 145–181.
2 Commission on Classification and Terminology of the International League Against Epilepsy (1989) Epilepsia 30 (4): 400–408.
3 Commission on Classification and Terminology of the International League Against Epilepsy (1981) Epilepsia 22: 489–501.
4 Commission on Classification and Terminology of the International League Against Epilepsy (1989) Epilepsia 30: 389–399.
5 Chadwick, D. (1990) Lancet 1114–1117.
6 Schmidt, D. (1991). In: Schmidt, D., Gram, L., Levy, R.H. (Eds.) Clinical Development of New Antiepileptic Drugs. Suppl. Epilepsy Research. Elsevier, Amsterdam (in press)
7 Leber, P.D. (1989) Epilepsia 30 (Suppl.): 57–63.
8 Dam, M. (1987). In: Dam, M. et al. (Eds.): Epilepsy: Progress in Treatment. Wiley & Sons. Chichester-New York-Brisbane-Toronto-Singapore. 257–261.
9 Heller, A.J., Chesterman, P., Elwes, R.D.C. et al. (1990). J. Neurol. 237 (Suppl.) 28–28.
10 Smith, D.B., Mattson, R.H., Cramer, J.A. et al (1987) Epilepsia 28, 3: 9–17.
11 Milligan, N., Richens, A. (1981) In: M. Alder and A. Richens (Eds.) Central Nervous System. Methods of Clinical Pharmacology 3. Mac Millan Press, New York, 139–152.
12 Jawad, S., Yuen, W.C., Oxley, J., Richens, A. (1986) Br. J. Clin. Pharmacol. 22: 191–193.
13 Yepez-Lasso, R., Duncan, J.S., Shorvon, S.D. (1990) Epilepsy Res. 5: 247–250
14 Schmidt, D., Leppik, I.E. (1988) Epilepsy Research, Suppl. I. Elsevier, Amsterdam.
15 Schmidt, D., Reininghaus, R., Winkel, R. (1988) In: Schmidt, D., Leppik, I.E. (Eds.) Compliance in Epilepsy. Epilepsy Research, Suppl 1. Elsevier, Amsterdam.
16 Sillanpää, M. (1990) Acta Paediat. Scand., Suppl. 368.
17 Gowers, W.R. (1885) Epilepsy and other chronic convul-

sive diseases. William Wood & Comp. New York.

18 Hakkarainen, H. (1981) Abstract, 14th International Epilepsy Symposium, Kyoto, Japan.

19 Mattson, R.H., Cramer, J.A., Collins, J.F. et al (1985) N. Engl. J. Med. 313: 145–151.

20 Schmidt, D. (1983) J. Neurol. Neurosurg. Psychatry 45: 1119–1124.

21 Schmidt, D. (1983) J. Neurol. 229: 221–226.

22 Schmidt, D., Richter, K. (1986) Ann. Neurol. 19: 85–87.

23 Temkin, R., Wilensky, A.J. (1986) Epilepsia 27: 644–645

24 Crawford, P., Chadwick, D. (1986) J. Neurol. Neurosurg. Psychiatry 49: 1251–1257.

25 Schmidt, D., Rohde, M., Wolf, P., Roeder-Wanner, U. (1986) Arch. Neurol. 43: 824–826.

26 Mumford, J.P., Dam, M. (1989) Br. J. Clin. Pharmacol. 27: 1015–1075.

27 Sander, J.W.A.S., Trevisol-Bittencourt, P.C., Hart, Y.M., Patsalos, P.N., Shorvon, S.D. (1990) Epilepsy Res. (in press).

28 Binnie, C.D., Debets, R.M.C., Engelsman, M. et al (1989) Epilepsy Res. 222–229.

29 Schmidt, D., Utech, K. (1986) Neurology 36: 217–221.

New Antiepileptic Drugs (Epilepsy Res. Suppl. 3)
Editors: F. Pisani, E. Perucca, G. Avanzini, A. Richens
© 1991 Elsevier Science Publishers B.V. (Biomedical Division)

CHAPTER 9

Pharmacokinetics and interactions of the new antiepileptic drugs

Emilio Perucca[1], Francesco Pisani[2]

[1]*Department of Pharmacology and Therapeutics, University of Pavia;* [2]*Neurological Clinic, University of Messina, Italy*

Introduction

More than 20 years have elapsed since sodium valproate, the latest major conventional antiepileptic agent, was introduced in clinical practice. This stagnation in drug development, however, has not prevented major advances in the pharmacological treatment of epilepsy, which occurred largely as a result of better use of available drugs. Advances in knowledge of clinical pharmacokinetics have been crucial in this process.

In recent years, a number of new anticonvulsants have entered pre-clinical and clinical development and a few of these are already at or near the marketing stage. Some of these agents derive from structural modifications of compounds already available, while others, which appear to be more innovative, were designed to interact specifically with excitatory or inhibitory pathways known to influence epileptogenesis in animal models. Experience with conventional agents and tighter regulatory guidelines have stimulated early characterization of the clinical pharmacokinetics of these new drugs, as well as research into their pharmacokinetic interactions with other agents prescribed in combination.

When assessing the therapeutic potential of the newer anticonvulsants, it is clear that consideration should also be given to their corresponding pharmacokinetic profile, which may result in clinically significant advantages (or disadvantages) with respect to conventional agents (Table 1).

Examination of the properties listed in Table 1 clearly reveals that the pharmacokinetic properties of most of the conventional antiepileptic drugs are far from ideal. Phenytoin, for example, shows non-linear (saturation) kinetics which complicates considerably the process of dosage titration in clinical practice. Additional pharmacokinetic problems with phenytoin include incomplete oral bioavailability (with some formulations), extensive plasma protein binding and involvement in a large number of clinically relevant drug interactions. Carbamazepine, in turn, exhibits an undesirably short half-life, dose-dependent autoinduction and a considerable interaction potential. Drug interactions are frequently observed also with valproic acid (a displacing agent and a metabolic inhibitor) and with phenobarbitone and primidone, both of which are potent inducers of the hepatic microsomal enzymes. Based on this background, it is clear that the pharmacokinetic profile is one feature for which there is room for improvement when designing new antiepileptic drugs.

TABLE 1

PHARMACOKINETIC PROPERTIES OF THE IDEAL ANTIEPILEPTIC DRUG

1.	Consistent and preferably complete oral availability;
2.	No or negligible plasma protein binding;
3.	Small interindividual variability in blood clearance (this property is more common for drugs which are not extensively metabolized);
4.	No active metabolites (unless bioactivation confers specific advantages);
5.	Half-life compatible with once or twice daily dosing but not too long to delay excessively attainment of steady-state (for reversibly acting drugs, half-lives of 15 to 50 h would be desirable);
6.	Good correlation between plasma drug level and pharmacological effect (to facilitate therapeutic drug monitoring);
7.	Lack of enzyme inducing or enzyme inhibiting activity;
8.	Pharmacokinetic parameters not liable to be altered by concomitantly given drugs

The purpose of the present article is to describe the pharmacokinetics and interactions of the most recently developed anticonvulsants in man. Particular emphasis will be placed on those drugs which are in the most advanced stages of development, but concise information will be provided also for some agents which have undergone only limited clinical testing.

Eterobarbital

Eterobarbital is a barbiturate derivative which in animal models exhibits a more favourable therapeutic ratio compared with phenobarbitone. Early clinical studies suggested that the drug retains a degree of efficacy similar to phenobarbitone but with lower sedating effects. The enthusiasm which followed these preliminary trials declined after the finding that no unchanged eterobarbital can be detected in blood and that the drug is extensively converted to phenobarbitone itself. In spite of this, it has been reported recently that subjects taking eterobarbital can tolerate relatively well serum levels of phenobarbitone which are normally associated with toxicity during phenobarbitone treatment[1]. To explain this observation, it has been speculated that an additional metabolite of eterobarbital (possibly the active momomethoxymethyl derivative, MMP)

could counteract the sedative effects of phenobarbitone.

After intravenous dosing, eterobarbital disappears rapidly from plasma, with a half-life of 10 to 100 minutes[2]. The pharmacokinetics of the drug after oral dosing in normal subjects has been investigated recently[3]. After ingestion of a single 400 mg dose of eterobarbital, no unchanged drug could be detected in serum. MMP appeared rapidly in the circulation but its concentration remained generally low and declined below the limit of detection (0.5 μg/ml) before 9.5 h in most subjects. The serum levels of the other metabolite phenobarbitone rose slowly and reached a peak between 24 and 48 h. Similar findings have been reported in epileptic patients[4].

Little information is available on drug interactions with eterobarbital. Since the drug is extensively converted to phenobarbitone, the interactions described for the latter drug should also apply to eterobarbital.

Felbamate

Felbamate is a promising new anticonvulsant currently undergoing clinical trials.

After oral administration, felbamate is readily absorbed from the gastrointestinal tract. The binding of the drug to plasma proteins is 24 to

35%[5]. The major metabolic pathway of felbamate involves hydroxylation and conjugation[5].

In normal volunteers, the half-life of felbamate after both single and multiple doses is about 20 h[6]. Wilensky et al[7] investigated the pharmacokinetics of felbamate in 8 adult epileptic patients treated with phenytoin or carbamazepine. After a single 200 mg oral dose, peak serum felbamate levels of 2.6–4.1 µg/ml were achieved in 1 to 4 h. Half-lives ranged from 11.2 to 16.1 h, clearance values from 34 to 65 ml/h/kg and apparent volumes of distribution from 0.7 to 1.0 l/kg. The single-dose study was followed by a 27-day treatment period during which felbamate dosage was increased stepwise from a minimum of 200 mg b.i.d. up to a maximum of 800 mg b.i.d. The serum levels of the drug at steady-state were linearly related to dose, although a trend towards a disproportionate increase of serum levels at the highest dosage was observed. Pharmacokinetic parameters were calculated again at the end of chronic dosing and showed no change in volume of distribution and a slight to moderate decrease in clearance and increase in half-life. Urinary excretion of unchanged drug accounted for 14 to 29% of the administered dose. In other studies, non linear pharmacokinetics have been observed at doses above 1600 mg/day.

In clinical trials in which felbamate was added to preexisting anticonvulsant therapy, the drug was found to increase serum phenytoin levels and decrease serum carbamazepine[5–9].

Gabapentin

Gabapentin is a compound structurally related to GABA, which has been shown to possess anticonvulsant activity in animal models and in seizure patients.

Human pharmacokinetic studies have demonstrated that gabapentin is rapidly absorbed from the gastrointestinal tract, peak serum levels being usually attained within 3 h. The absolute oral availability is at least 60%[10]. The drug is not bound to plasma proteins and has a volume of distribution at steady-state of about 50 l. A cortex/serum partition ratio of 0.80 has been reported in one patient undergoing neurosurgery[11]. Gabapentin is not metabolized and is eliminated unchanged by the kidney[12,13]. After oral administration, unchanged drug in urine accounts for 75 to 81% of the dose, the remainder being recovered in the faeces[12].

In subjects with normal renal function, the elimination half-life of gabapentin is about 6 h while the renal clearance is 120 to 160 ml/min. A remarkable aspect of gabapentin kinetics appears to be a very small degree of intra- and interindividual variability in absorption and disposition parameters[14]. Administration of doses ranging between 100 and 900 mg have shown that the disposition of the drug is independent of the dose, whereas gastrointestinal absorption appears to be reduced at the highest dosage. After multiple dosing the pharmacokinetics of the drug are similar to those observed after a single dose.

Gabapentin is devoid of enzyme inducing properties. As expected from the absence of plasma protein binding and the exclusively renal route of elimination, gabapentin does not appear to be involved in significant pharmacokinetic interactions with concurrently prescribed anticonvulsants[10].

Lamotrigine

Lamotrigine is a new anticonvulsant which in animal models shows a spectrum of activity similar to that of phenytoin. It has been suggested that the mode of action of lamotrigine involves inhibition of the release of the excitatory transmitter glutamate. Its antiepileptic efficacy has been confirmed in clinical trials.

Lamotrigine is rapidly absorbed from the

gastrointestinal tract, peak serum levels being usually attained within 3 h of drug intake. A study comparing intravenous with oral administration found that the absolute oral bioavailability of lamotrigine is virtually complete[15].

The apparent volume of distribution of lamotrigine is about 1.1 l/kg[15]. Binding to plasma proteins is about 55%. Saliva concentrations are approximately 45% of the plasma concentrations[16].

After intake of single doses by normal volunteers, the elimination half-life of lamotrigine is about 25 h (range 14 to 50 h)[15,16]. Clearance is 0.55 ml/min/kg. No evidence of dose-dependent kinetics has been found, at least within the 7.5–240 mg range. Pharmacokinetic parameters after multiple dosing agree closely with those predicted from single dose experiments, suggesting the absence of autoinduction of metabolism[16]. Lamotrigine is extensively metabolized by conjugation. A glucuronide conjugate accounts for about 90% of the total urinary recovery which, in turn, accounts for about 70% of the administered dose[15]. The elimination of lamotrigine is moderately reduced in patients with Gilbert's syndrome, as one might anticipate, since the drug is eliminated mainly by glucuronide conjugation[17].

Lamotrigine is devoid of enzyme inducing activity and does not appear to affect significantly the kinetics of concurrently administered anticonvulsants[18]. On the other hand, concurrent anticonvulsant therapy has a marked influence on lamotrigine pharmacokinetics. The metabolism of lamotrigine is markedly accelerated in patients receiving enzyme-inducing anticonvulsants such as phenytoin, phenobarbitone, primidone and carbamazepine. The half life of the drug in these patients is about 15 h (range 8 to 33 h) compared with 25 h in normal subjects[19,20]. Conversely, lamotrigine metabolism is inhibited by valproic acid: valproate-treated patients show lamotrigine half-lives in the 30 to 90 h range (mean 60 h). Patients receiving valproate together with en-

zyme-inducing drugs exhibit a half-life of about 30 h on average[21]. These interactions are clinically relevant, different lamotrigine dosages being indicated depending on the presence and type of associated drug therapy.

Loreclezole

Loreclezole is a triazole derivative structurally unrelated to other anticonvulsants. The drug is currently undergoing initial clinical testing.

After oral dosing. loreclezole is absorbed rapidly from the gastrointestinal tract. In patients receiving concurrent treatment with conventional antiepileptic agents the serum levels of loreclezole are several-fold lower than those observed in normal volunteers receiving equivalent dosages[22]. Loreclezole has an extremely long half-life. In patients receiving combination therapy with other anticonvulsants loreclezole half-lives range from 10 to 34 hours.

Losigamone

Losigamone is a recently developed anticonvulsant whose chemical structure resembles that of a class of naturally occurring compounds.

Given to normal volunteers, losigamone is absorbed rapidly from the gastrointestinal tract, peak plasma levels being linearly correlated with dosage. The terminal half-life is 4 to 5 h[23]. No information on drug interactions is available to date.

Oxcarbazepine

Oxcarbazepine is a keto-derivative of carbamazepine which exerts antiepileptic effects similar to those of carbamazepine, possibly associated with improved clinical tolerability.

Oxcarbazepine is readily absorbed from the gastrointestinal tract. After administration of a single oral dose (600 or 900 mg), peak plasma oxcarbazepine levels occur within 1 h but they are very low (about 4 μmol/l) and decline rapidly with half-lives of 1.0 to 2.5 h[24,25]. These data suggest that orally administered oxcarbazepine may be incompletely bioavailable due to substantial pre-systemic metabolic clearance.

The drug is extensively metabolized. Studies based on the analysis of plasma and urinary levels of unchanged drug and metabolites suggest that the major metabolic pathway involves reduction to 10-hydroxy-oxcarbazepine (an active metabolite), which exists in two enantiomeric forms possessing similar anticonvulsant activity in animal models[24,26]. Other minor metabolites, including the O-sulphate and O-glucuronide, have also been identified in the urine of man[26]. Unlike carbamazepine, the metabolic pattern of oxcarbazepine does not involve formation of an epoxide intermediate.

The pharmacokinetic profile of oxcarbazepine and the pharmacological profile of its metabolites indicates that the compound can be considered largely as a pro-drug for 10-hydroxy-carbazepine[27]. After single oral oxcarbazepine doses, peak plasma levels of this active metabolite occur at about 8 h[28]. In chronically treated patients, the 10-hydroxy-metabolite is found in serum at concentrations very much higher than those of the parent drug, confirming that this compound is primarily responsible for the observed pharmacological activity[29]. The relationship between oxcarbazepine dosage and plasma 10-hydroxy-carbazepine levels appears to be linear[25]. The terminal half-life of this active metabolite is about 10 h after both single and multiple doses[24,25,30]. The metabolite is eliminated mainly by conjugation with glucuronic acid. A minor proportion, however, is excreted unchanged in urine or transformed to hydroxylated products which include the 10,11-trans-diol, also formed after administration of carbamazepine[26]. A recent case report indicated that 10-hydroxy-carbazepine crosses the human placenta and is found in neonatal serum at concentrations similar to those found in the mother[31].

Oxcarbazepine seems to have a low interaction potential. Its pharmacokinetics in monotherapy patients have been reported to be similar to those observed in patients receiving other anticonvulsants in combination[27]. Unlike carbamazepine, oxcarbazepine has little or no hepatic microsomal enzyme inducing properties in man and therefore it is less likely to affect the kinetics of other concurrently given drugs[29,30].

Ralitoline

Ralitoline is a structurally novel anticonvulsant which has undergone initial clinical testing.

The drug is well absorbed from the gastrointestinal tract, 90% of a radioactive dose being recovered in urine within 72 h. Peak serum levels are usually attained within 4 h after a single dose[32].

Ralitoline is moderately (74%) bound to plasma proteins. The elimination half-life is about 3 to 5 hours, while the mean transit time is approximately 6 h. Ralitoline shows linear pharmacokinetics over a wide dose range. The drug is extensively metabolized, less than 1% of the dose being excreted in urine as unchanged substance. Three oxidation products have been identified in the urine of man. Only the unchanged drug is responsible for the pharmacological activity[32].

Concurrent administration of phenytoin or phenobarbitone decreases serum ralitoline levels and shortens ralitoline half-life, presumably due to enzyme induction[33]. Conversely, ralitoline does not appear to alter the serum levels of phenytoin or phenobarbitone at steady-state[33].

Stiripentol

Stiripentol has been considered a promising new drug in open clinical trials, although the occurrence of important pharmacokinetic interactions with other drugs complicates evaluation of its therapeutic potential.

The pharmacokinetics of stiripentol have been extensively studied in non-human primates [34] and in man[35,36]. The drug appears to be absorbed readily from the gastrointestinal tract and to be extensively (>90%) bound to plasma proteins. Stiripentol is eliminated by metabolism through conjugation, opening of the methylene-dioxy ring, and other pathways. It has now been clearly demonstrated that the biotransformation of the drug is saturable within the clinically used dosage range, resulting in non-linear (Michaelis-Menten type) kinetics.

Evidence of non-linear stiripentol kinetics was initially provided by Levy and coworkers[37], who observed an almost 8-fold decrease in oral clearance during 8 days of multiple dosing in normal volunteers. Dose-dependent changes in stiripentol clearance in normal volunteers were subsequently demonstrated: mean clearance values in 6 subjects were found to decrease from 479 l/day at a daily dosage of 600 mg to 224 l/day at a daily dosage of 1800 mg[35]. A similar study in epileptic patients receiving concomitant anticonvulsant drug treatment revealed mean clearance values of 2607 l/day at 600 mg daily, 1254 l/day at 1200 mg daily and 536 l/day at 2400 mg/daily[36]: although these findings confirm the occurrence of dose-dependent kinetics, it is clear that at comparable dosages clearance values were higher in patients than in normal volunteers, presumably due to induction of stiripentol metabolism by concurrently administered anticonvulsants. Because of these pharmacokinetic properties, serum stiripentol levels tend to increase disproportionately after a dosage increment, in a way similar to that

observed with phenytoin. Under these conditions, careful individualization in dosage, possibly assisted by serum drug level monitoring, becomes mandatory.

The interactions of stiripentol with other anti-epileptic drugs are complex. As reported above, enzyme inducing anticonvulsants stimulate stiripentol metabolism and cause a marked decrease of serum stiripentol levels at steady-state. Conversely, stiripentol has been shown to exert prominent effects on the disposition of other anticonvulsants. In particular, stiripentol inhibits markedly phenytoin and carbamazepine metabolism: after adding stiripentol, phenytoin clearance decreased in one study from 29.5 ± 13.4 to 6.5 ± 2.6 l/kg while carbamazepine clearance decreased from 6.1 ± 1.1 to 2.0 ± 0.7 l/kg[36]. If the dosages of phenytoin and carbamazepine are not reduced, this interaction may result in serious toxicity. Stiripentol has also been shown to inhibit the metabolism of phenobarbitone, primidone and, to a lesser extent, valproic acid[38,39]. In view of the obvious clinical relevance of most of these interactions, careful monitoring of the serum levels of associated anticonvulsants is essential when stiripentol is used as add-on therapy.

Tiagabine

Tiagabine is a novel anticonvulsant whose mode of action is considered to be mediated by inhibition of GABA reuptake from the synaptic cleft.

After oral administration of single doses ranging from 2 to 24 mg to normal volunteers, the absorption of tiagabine is very rapid and peak plasma levels are usually reached within 1 hour. Apparent half-life values range from 4.5 to 13.4 hours[40]. No evidence of dose-dependent kinetics has been found within the indicated dose range, and no changes in pharmacokinetic parameters have been observed following administration of doses up to 10 mg daily for five consecutive days[40].

Topiramate

Topiramate is a new anticonvulsant drug which has shown promising activity in animal models and has undergone initial clinical testing in Europe and the United States.

After oral administration of single doses (100 to 1200 mg) to normal volunteers, peak plasma topiramate levels are linearly related to dose and occur after 1.8 to 4.3 h. The apparent volume of distribution, calculated by assuming complete oral availability, is 0.6–0.8 l/kg. Half-life values of 19 to 23 h and clearance values of 21 to 36 ml/min have been reported[41]. Similar pharmacokinetic parameters, irrespective of the dose used, have been described after multiple dosing. Topiramate is excreted unchanged in urine to an important extent, since renal clearance accounts for about 60% of total plasma clearance[42].

Preliminary observations suggest that topiramate does not alter the serum concentrations of concurrently administered phenytoin or carbamazepine[42].

Vigabatrin

Vigabatrin (gamma-vinyl-GABA) results from a systematic search into possible ways of increasing GABA-ergic inhibition through interference with GABA metabolism. The drug is a specific, irreversible, enzyme-activated inhibitor of GABA-transaminase and increases brain GABA levels in animals and in man[43]. Vigabatrin exists in two allosteric forms, only the S(+)-enantiomer being pharmacologically active. Its short- and long-term efficacy in epileptic patients is well documented[44–46].

Vigabatrin is absorbed rapidly and almost completely from the gastrointestinal tract, peak plasma levels being usually attained within the first 2 hours irrespective of whether the drug is taken with food or in the fasting state[43,47]. Pharmacokinetic studies using a stereoselective assay have shown that at the time of peak the concentration of the R(−)-enantiomer is about twice that of the active S(+)-enantiomer[48–50].

The apparent volume of distribution of vigabatrin in adults is estimated to be about 0.8 l/kg. Larger volumes of distribution have been described in children[50].

Vigabatrin is excreted to a large extent by the kidney, the renal clearance of unchanged drug accounting for about 60–70% of the total clearance. Urinary excretion of racemic drug during the first 24 h after a single dose accounts for 62–68% of the administered dose[43]. Although the renal clearance of the two enantiomers is similar, the urinary recovery after a single dose is about 50% for the S(+)-enantiomer and 65% for the S(−)-enantiomer[48].

In adults, the half-life of vigabatrin in patients with normal renal function ranges between 5 and 11 h, without significant differences between the two enantiomers[48]. Comparable values have been observed in children in whom, however, the half-life of the R(−) enantiomer is shorter than that of the other enantiomer[50]. A prolongation of the half-life has been reported in the elderly, the magnitude of the change being dependent on the decrease in creatinine clearance[49]. Due to the irreversible mode of action of the drug, the plasma half-life bears practically no relationship to the duration of pharmacological effect.

In adults, the total plasma clearance is about 1.7 ml/min/kg; larger clearance values have been reported in children[50], whereas a decrease in clearance has been described in elderly subjects with decreased glomerular filtration rate[49]. Linearity of vigabatrin kinetics has been demonstrated within the 1 to 3 g dose range[43].

The renal clearance of vigabatrin is reduced in

the presence of impaired renal function. In elderly patients with reduced glomerular filtration rate the decrease in vigabatrin clearance and the prolongation in vigabatrin half-life are nonlinearly related with the reduction in creatinine clearance[49].

There is no evidence that concomitantly prescribed antiepileptic drugs can modify the pharmacokinetics of vigabatrin. Likewise, vigabatrin does not affect the plasma levels of concurrently administered carbamazepine and valproic acid[51–53]. A reduction in plasma levels of phenobarbitone and primidone after addition of vigabatrin has been reported[54], but does not appear to occur consistently[51–53].

By far the most significant pharmacokinetic drug interaction described to date is a lowering effect of vigabatrin on the plasma levels of concurrently administered phenytoin[51–54]. The fall in plasma phenytoin averages 20 to 30% of the baseline level, but the interindividual variability in the degree of interaction is considerable. Part of this variability may be related to time, since it has been shown that the decrease in plasma phenytoin may not occur until more than 4 weeks have elapsed after starting vigabatrin[55]. The mechanism of this interaction is unknown. It has been shown, in any case, that it cannot be ascribed to displacement of phenytoin from plasma proteins[54,55]. Stimulation of phenytoin metabolism by vigabatrin is also unlikely, since there is no evidence of vigabatrin acting as an enzyme inducer. It can be speculated that vigabatrin may decrease the absorption of phenytoin, but further studies are required to confirm this.

Zonisamide

Zonisamide is a relatively weak inhibitor of carbonic anhydrase which in animal models displays anticonvulsant effects similar to those of phenytoin and carbamazepine. Clinical efficacy in human epilepsy has been demonstrated in a number of trials.

Zonisamide is well absorbed from the gastrointestinal tract, peak plasma levels being observed within 4–7 hours after oral intake[56,57]. In blood, the drug accumulates extensively within erythrocytes (probably due to binding by red cell carbonic anhydrase) and is approximately 50% bound to plasma proteins[56,57]. Binding to plasma proteins is a saturable process and decreases with increasing drug level within the clinically occurring concentration range[56,58].

The plasma elimination half-life of zonisamide (estimated with some approximation since the kinetics of the drug are nonlinear) has been reported to be about 60 hours in normal subjects, about 36 hours in carbamazepine-treated patients and about 27 hours in phenytoin-treated patients, suggesting that the metabolism of zonisamide is induced by other anticonvulsants[59–61]. Half-life values calculated from red cell concentrations are longer than the half-lives in plasma[61]. Zonisamide is partly excreted unchanged in urine and partly metabolized via glucuronic acid conjugation, N-O cleavage and other pathways[57]. The proportion of the dosage excreted unchanged ranges from 8 to 50%[56,57]. The metabolic pathways of zonisamide may become saturated within the clinically occurring concentration range, resulting in a nonlinear relationship between plasma levels and dosage[58–60,62]. Non-linear kinetics may also explain the observation that zonisamide levels at steady state are higher than those predicted from single-dose kinetics[59–62].

It has been reported that zonisamide may increase the serum levels of concurrently administered carbamazepine, possibly by interfering with its metabolism[62].

References

1 Smith, D.B., Goldstein, S.G., Roomet, A. (1986) Epi-

lepsia 27: 149–155.

2 Matsumoto, H., Gallagher, B.B. (1976) In: Janz, D. (ed.), Epileptology. George Thieme, Stuttgart, pp. 122–129.

3 Barzaghi, N., Gatti, G., Manni, R., Galimberti, C.A., Zucca, C., Perucca, E., Tartara, A. (1991) Biol. Neurol. (in press).

4 Goldberg, M., Gal, J., Cho, A.K., Henden, D.J. (1976) Ann. Neurol. 5: 121–126.

5 Fuerst, R.H., Graves, N.M., Leppik, I.E., Brundage, R.C., Holmes, G.B., Remmel, R.P. (1988) Epilepsia 29: 488–491.

6 Leppik, I., Graves, N.M. (1989) In: Levy, R.H., Dreifuss, F.E., Mattson, R.H., Meldrum, B.S., Penry, J.K. (eds.), Antiepileptic Drugs. Raven Press, New York, pp. 983–990.

7 Wilensky, A.J., Friel, P.N., Ojemann, L.M., Kupferberg, H.S., Levy, R.H. (1986) Epilepsia 26: 602–606.

8 Holmes, G.B., Graves, N.M., Leppik, I.E., Fuerts, R.H. (1987) Epilepsia 28: 578–579.

9 Leppik, I. (1989) In: Pitlick, W.H. (ed.), Antiepileptic Drug Interactions. Demos, New York, pp. 115–128.

10 Schmidt, B. (1989) In: Levy, R.H., Dreifuss, F.E., Mattson, R.H., Meldrum, B.S., Penry, J.K. (eds.), Antiepileptic Drugs. Raven Press, New York, pp. 925–935.

11 Ojemann, L.M., Friel, P.N., Ojemann, G.A. (1988) Epilepsia 29: 694.

12 Vollmer, K.O., Von Hodenberg, A., Kolle, E.U. (1986) Arzneim. Forsch. 36: 830–839.

13 Ben Menachem, E., Persson, L.I. (1989) 18th International Epilepsy Congress, New Delhi, Abstract 137.

14 Voller, K.O., Kolle, E.U., Ekstedt, J., Forsgren, L. (1986) Acta Pharmacol. Toxicol. (Suppl. 5).

15 Yuen, W.C., Peck, A.W. (1987) Epilepsia 28: 582.

16 Cohen, A.F., Land, G.S., Breimer, D.D., Yuen, W.C., Winton, C., Peck, A.W. (1987) Clin. Pharmacol. Ther. 42: 535–541.

17 Cohen, A.F., Posner, J., Land, G., Winton, C. (1986) II World Conference on Clinical Pharmacology and Therapeutics, Stockholm, Abstract.

18 Wolf, P. 18th International Epilepsy Congress, New Delhi, 1989, Abstract 57.

19 Binnie, C.D., Van Emde Boas, W., Kastelejin-Nolste-Trenite, D.G.A. et al (1986) Epilepsia 27: 248–254.

20 Lai, A., Wargin, W.A., Garnett, W.R., Ramsay, R.E., Pellock, J.M., Vasquez, D., Hubbell, J.P. (1986) Epilepsia 27: 647–648.

21 Jawad, C.D., Yuen, W.C., Peck, A.W., Hamilton, M.J., Oxley, J.R., Richens, A. (1987) Epilepsy Res. 1: 194–201.

22 De Beukelaar, F., Tritsmans, L. (1991) This volume.

23 Stein, U., Klessing, K., Chatterjee, S.S. (1991) This volume.

24 Theisohn, M., Heimann, G. (1982) Eur. J. Clin. Pharmacol. 22: 545–551.

25 Dickinson, R.G., Hooper, W.D., Dunstan, P.R., Eadie, M.J. (1989) Eur. J. Clin. Pharmacol. 37: 69–74.

26 Schutz, H., Feldmann, K.F., Faigle, J.W., Kriemler, H.P., Winkler, T. (1986) Xenobiotica 16: 769–778.

27 Faigle, J.W., Menge, G.P. (1990) Int. Clin. Psychopharmacol. 5 (suppl. 1): 73–82.

28 Kristensen, O., Klitgaraard, N.A., Jonsson, B., Sindrup, S. (1983) Acta Neurol. Scand. 68: 145–150.

29 Anonymous (1989) Lancet 2: 196–198.

30 Brodie, M.J., Larkin, J.G., McKee, P.J., Forrest, G., Scobie, G., Beastall, G.H., Lowrie, J.I. (1989) 18th International Epilepsy Congress, New Delhi, Abstract 105.

31 Bulau, P., Paar, W.D., Von Unruh, G.E. (1988) Eur. J. Clin. Pharmacol. 34: 311–313.

32 Von Hodenberg, A., Katzinski, L., Fecht-Kempter, I., Vollmer, K.O. (1987) In: Aiache, J.M., Hirtz, J. (eds.), Proceed. Third Eur. Congress Biopharmac. Pharmacokin., Vol. 3, pp. 460–470.

33 Von Hodenberg, A., Fecht-Kempter, I., Vollmer, K.O., Schleyer, I., Bockbrader, H.N., Sedman, A. (1989) Eur. J. Clin. Pharmacol. 36 (Suppl.): A270.

34 Lin, H.S., Levy, R.H. (1983) Epilepsia 24: 692–702.

35 Levy, R.H., Loiseau, P., Guyot, M., Blehaut, H., Tor, J., Moreland, T.A. (1984) Epilepsia 25: 486–491.

36 Levy, R.H., Loiseau, P., Guyot, M., Blehaut, H., Tor, J., Moreland, T.A. (1984) Clin. Pharmacol. Ther. 36: 661–669.

37 Levy, R.H., Lin, H.S., Blehaut, H.M., Tor, J.A. (1983) J. Clin. Pharmacol. 23: 523–533.

38 Loiseau, P., Guyot, A.M., Acheampong, A., Tor, J. (1988) Epilepsia 29: 709.

39 Loiseau, P., Levy, R.H., Tor, J. (1989) In: Pitlick, W.H. (ed.), Antiepileptic Drug Interactions. Demos, New York, pp. 129–142.

40 Pierce, M.W., Suzdak, P.D., Gustavson, L.E., Mengel, H.B., McKelvy, J.F., Mant, T. (1991) This volume.

41 Easterling, D.E., Zakszewski, T., Moyer, M.D., Margul, B.L., Marriott, T.B., Nayak, R.K. (1988) Epilepsia 29: 662.

42 Doose, D.R., Scott, V.V., Margul, B.L., Marriott, T.B., Nayak, R.K. (1988) Epilepsia 29: 662.

43 Schechter, P.J. (1989) Brit J Clin Pharmacol 27: 195–225.

44 Mumford, J.P., Dam, M. (1989) Br. J. Clin. Pharmacol. 27: 101S –107S.

45 Tartara, A., Manni, R., Galimberti, C.A., Mumford, J.P., Iudice, A., Perucca, E. (1989) J. Neurol. Neurosurg. Psychiatry 52: 467–471.

46 Wilder, B.J., Penry, J.K., Ranger, R.J., Riela, A., Margul, B.L., Marriott, T.B. (1988) Epilepsia 29: 698.

47 Frisk-Holmberg, M., Kerth, P., Meyer, P. (1989) Br. J. Clin. Pharmacol. 27: 235–255.

48 Haegele, K.D., Schechter, P.J. (1986) Clin. Pharmacol. Ther. 40: 581–586.

49 Haegele, K., Huebert, N.D., Ebel, M., Tell, G., Schechter, P.J. (1988) Clin. Pharmacol. Ther. 44: 558–565.

50 Rey, E., Pons, G., Richard, M.O., Vauzelle, F., Chiron, C.C., Dulac, O., D'Athis, P., Beaumont, D., Olive, G. (1989) 18th International Epilepsy Congress, New Delhi, Abstract 559.

51 Rimmer, E.M., Richens, A. (1984) Lancet i: 189–190.

52 Tartara, A., Manni, R., Galimberti, C.A., Hardenberg, J., Orwin, J., Perucca, E. (1986) Epilepsia 26: 713–723.

53 Tassinari, C.A., Michelucci, R., Ambrosetto, G., Salvi, F. (1987) Arch. Neurol. 44: 907–910.

54 Browne, T.R., Mattson, R.H., Penry, J.K., Smith, D.B. et al (1987) Neurology 37: 184–189.

55 Rimmer, E.M., Richens, A. (1989) Br. J. Clin. Pharmacol. 27: 27S –33S.

56 Matsumoto, K., Miyazaki, H., Fujii, T., Kagemoto, A., Maeda, T., Hashimoto, M. (1983) Arznm. Forsch. 33: 96–98.

57 Ito, T., Yamaguchi, T., Miyazaki, H. (1982) Arznm. Forsch 32: 1581–1586.

58 Sanchez, R.M., Marcos, J.L., Ramsay, R.E., Vasquez, D., Guterman, A., Riddle, G., McJilton, J., Wilder, B.J., Sackellares, C., Browne, T. (1986) Epilepsia 27: 647.

59 Sackellares, J.C., Donofrio, P.D., Wagner, J.G., Abou-Khalil, B., Berent, S., Aasved-Hoyt, K. (1985) Epilepsia 26: 206–211.

60 Wagner, J.G., Sackellares, J.C., Donofrio, P.D., Berent, S., Sakmar, E. (1984) Ther. Drug Monit. 6: 277–283.

61 Ojemann, L.M., Shastri, R.A., Wilensky, A.J., Friel, P.N., Levy, R.H., McLean, J.R., Buchanan, R.A. (1986) Ther. Drug Monit. 8: 293–296.

62 Wilensky, A.J., Friel, P.N., Ojemann, L.M., Dodrill, C.B., McCormick, K.B., Levy, R.H. (1985) Epilepsia 26: 212–220.

New Antiepileptic Drugs (Epilepsy Res. Suppl. 3)
Editors: F. Pisani, E. Perucca, G. Avanzini, A. Richens
© 1991 Elsevier Science Publishers B.V. (Biomedical Division)

CHAPTER 10

The efficacy and safety of new antiepileptic drugs

Alan Richens

Department of Pharmacology and Therapeutics, University of Wales College of Medicine, Heath Park, Cardiff CF4 4XN, UK

The traditional antiepileptic drugs such as phenobarbitone, phenytoin, primidone and carbamazepine are effective forms of treatment in a substantial proportion of patients with partial and generalised seizures. However, they have a number of drawbacks which have limited their value and have sometimes made drug therapy a heavy burden for the patient. Table 1 summarises some of these problems.

It is general experience that 20–30 percent of patients who develop partial and secondarily generalised seizures do not have their seizures adequately controlled by drug therapy and therefore physicians who specialise in the field of epilepsy have large numbers of such patients waiting for a new and more effective remedy to appear. The problems of drug resistance are considered at length elsewhere in this volume.

TABLE 1

PROBLEMS WITH OLD DRUGS

Incomplete efficacy
Narrow therapeutic ratio
CNS depressant effects
Many idiosyncratic adverse effects
Tolerance
Enzyme induction
Interactions
Difficult kinetics

The therapeutic effectiveness of the older antiepileptic drugs has usually been limited by their narrow therapeutic ratio, i.e. the ratio of toxic dose against the effective dose. The acute cerebellar toxicity of phenytoin has long been recognised but remains one of the most common and dose-limiting adverse effects of the drug. Similar symptoms can be produced by carbamazepine although usually less florid in nature. Phenobarbitone and primidone have the particular disadvantage of causing sedation, impaired concentration and learning and behavioural adverse effects in children. Although it is optimistic to expect that a drug with a therapeutic effect on the central nervous system will produce no central adverse effects, it is hoped that an improved therapeutic ratio may be seen with some of the novel compounds currently being developed.

In addition to predictable adverse reactions, many of the standard antiepileptic drugs are responsible for a number of idiosyncratic effects i.e. effects which would not be predicted from their known pharmacological actions. Such examples are gingival hyperplasia caused by phenytoin and hepatotoxicity caused by valproate. The mechanism of these idiosyncratic effects is totally unrelated to their therapeutic effects as far as we know. So often new drugs founder on these unexpected effects, especially if they occur in man without having been seen in animal screening

90

tests. The latter, however, are usually not designed to pick up unusual reactions because of their very nature. It would be nice if we could anticipate new drugs without idiosyncratic effects but it is difficult to design a drug to fulfil this requirement.

Tolerance is a recognised problem with both benzodiazepine and barbiturate drugs and limits their value in clinical practice. Apart from the loss of effect which results with continuing therapy, rebound effects may complicate withdrawal of the treatment. The mechanism of this tolerance is central, due to a down-regulation of receptor sites, but in the case of barbiturates, auto induction of metabolism also contributes. The latter effect is seen also with carbamazepine and may necessitate an increase in dose to compensate for a fall in plasma level.

Hepatic enzyme induction is a feature of phenytoin and primidone treatment also. It is well recognised that this phenomenon can lead to metabolic disturbances such as anticonvulsant osteomalacia and folate deficiency, as well as causing drug interactions. The latter can reduce the contraceptive efficiency of the pill and give rise to difficulties in anticoagulant control with warfarin, as well as many other interactions.

Enzyme inhibition is another cause of drug interactions during antiepileptic drug therapy. Sodium valproate, for instance, will inhibit the metabolism of carbamazepine, and the related valpromide is even more potent in this respect. Sulthiame causes a marked inhibition of phenytoin metabolism, frequently giving rise to phenytoin intoxication during combination therapy, and this has contributed to the decline in the use of sulthiame.

A final problem with some of the conventional antiepileptic drugs is their kinetic properties. Phenytoin, for example, undergoes saturable metabolism in the liver, such that increments in dose produce a much greater increase in plasma

level than would be predicted from the size of the increment. This phenomenon becomes more pronounced as the therapeutic range of plasma concentration is reached and intoxication can readily be provoked by apparently reasonable increases in dose. For this drug, therefore, plasma level monitoring is mandatory in optimising the dose. Phenytoin and valproate are extensively bound to plasma proteins and this can make them vulnerable to interaction with each other and with other drugs.

Conventional antiepileptic drugs therefore leave much to be desired as therapeutic agents. New drugs with an improved benefit/risk ratio are eagerly awaited.

What expectation do we have of new drugs?

In theory, antiepileptic drugs can benefit patients in four possible ways:
1. To prevent epilepsy developing in the first place. If the onset of seizures can be predicted, then prophylaxis may prevent the onset. This has been the reason for administering antiepileptic drugs to patients undergoing neurosurgery or after severe brain injury, particularly in those patients in whom a high risk is known to exist by the nature of the damage. The optimism behind this practice has, however, been doubted in recent studies and a decline in this type of therapy can be expected, not least because many of the patients receiving prophylactic therapy would never have needed it in the first place and are merely exposed to the risk of adverse effects. Experimental evidence with new therapeutic agents, however, suggests that this approach might be more fruitful in the future. Excessive excitatory amino acid transmission is thought to be an important factor in the pathophysiology of ischaemic brain damage, probably by producing calcium overload in cerebral neurones, and glutamate receptor antagonists have been shown to have an impressive cereb-

roprotective effect if given before or even during the few hours after the ischaemic injury. Clinical evidence is accumulating that this beneficial effect may not be confined only to experimental animals. A new therapeutic approach to the management of stroke and head injury can be anticipated and the benefits may include a reduced liability to seizures. Perhaps also the cerebral damage produced by status epilepticus may be lessened and even more excitingly, the Ammon's horn sclerosis resulting from prolonged febrile convulsions, might be prevented by the use of these agents. This would be real progress in the prevention of epilepsy. The scientific evidence supporting these hopes is discussed in the Chapter by Chapman and Meldrum.

2. To cure epilepsy once it has started. This is little more than a pipe dream at present, but experimental work suggests that an upregulation of glutamate receptors is a possible cause of seizure disorders. If this is so, it is conceivable that some way will be devised in future of permanently down-regulating the abnormal receptors, leading to a permanent cure.

3. To stop seizure activity and thereby to improve the prognosis. It has been suggested by a number of clinicians (see for example the Chapter by Reynolds) that early and effective control of seiz-

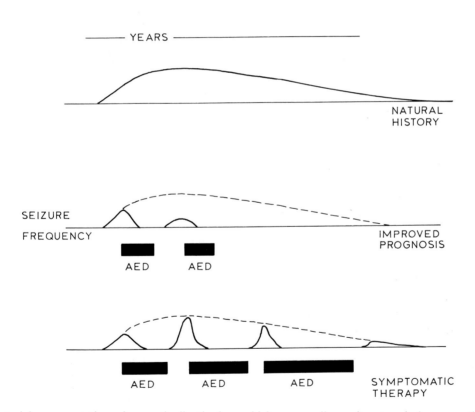

Fig 1. Pictorial representation of an antiepileptic drug which cures epilepsy (top trace), improves the prognosis (middle trace) or acts as symptomatic therapy only (bottom trace).

ures may alter the natural history of the disorder in the individual patient, a logical extension of the concept that seizures beget seizures. This idea is illustrated pictorially in Figure 1. Early treatment would reduce the liability to further seizures and shorten the overall duration of the disorder. Various arguments have been put forward to support this hypothesis. For instance, a number of studies have shown that the remission rate is inversely related to the number of seizures which occurred before treatment was instituted; it is suggested that once the seizure disorder establishes itself the more resistant it will be as a result of those seizures. The counter to this argument is that the epilepsy is simply expressing its natural history. Frequent seizures before presentation merely herald the onset of severe epilepsy. Attempts to test this hypothesis are just beginning in the form of randomising patients with one seizure to early treatment or to no treatment until the second seizure occurs.

4. To provide symptomatic therapy only. In this way, drug therapy simply suppresses seizures but does not alter the natural history of the condition (Figure 1). The frequency and/or the severity of seizures will be reduced but as soon as the treatment is withdrawn they will return to the level determined by the natural course of the epilepsy.

Whether conventional antiepileptic drug therapy is merely symptomatic therapy or additionally improves the prognosis, will be a matter for debate for some time to come. My belief is that they offer symptomatic relief only and sometimes do that rather ineffectively. However, the necessity for argument may well be overtaken by the introduction of new agents which have clearly been shown in animal studies and preferably in man also to offer benefits above and beyond symptomatic therapy.

What properties should new drugs have?

1. Efficacy

Table 2 lists some of the features that we should be looking for in new antiepileptic agents.

One of the first requirements is that the mode of action should be known. The days when new drugs are identified only on pharmacological screening tests are now over. Even if a new compound emerges in this way, it is mandatory that it is submitted to comprehensive pharmacological testing in order to identify its site and mode of action. This requires considerable resources but few pharmaceutical companies now remain that do not have these facilities available to them.

The reason for knowing the mode of action is that we will be able to predict better their likely usage, spectrum of activity and adverse effects. We may also be able to combine agents together to complement or enhance each others actions, such as inhibiting the breakdown of the inhibitory transmitter GABA with one drug while simultaneously inhibiting GABA reuptake into neurones and glial cells with another agent. It may also help us to avoid continuation of drugs whose actions are likely to conflict. Finally, it may tell us a great deal about the pathophysiology of the epilepsy itself.

In Table 2 I have suggested that the drug should

TABLE 2

EFFICACY – WHAT WE DO WANT

Known mode of action – specific action
Broad spectrum?
More efficacious than old drugs
No tolerance or withdrawal effects
Simple kinetics, simple dosing
No plasma level monitoring
Oral parenteral formulation
Once or twice daily dosing
No sustained release formulations
Inexpensive

be broad in its spectrum of activity although I accept that this is a point that could be debated. Using the simile of antibiotic therapy, microbiologists argue that narrow spectrum, targetted antimicrobial agents are the ideal but the use of these requires sophisticated diagnostic ability on the part of the clinician. Furthermore, the Industry has concentrated in recent years in broadening the spectrum of activity of their products rather than exploring the targetted approach. It can be argued that they are simply responding to a need – the clinician has adopted the sledge hammer approach to antibiotic therapy despite the bacteriologist's ideals. With antiepileptic drug therapy a broad spectrum agent would probably be of more use to us because different seizure types often coexist. Also the less sophisticated clinician will make fewer mistakes such as treating juvenile myoclonic epilepsy with carbamazepine or complex partial seizures with ethosuximide. I accept, however, that this view may need to be modified as our knowledge of the pathophysiology of epilepsy deepens.

That we require more efficacious compounds is self evident from the prevalence of drug resistant epilepsy. Drugs which will control complex partial seizures, Lennox-Gastaut syndrome or some of the other types of intractable epilepsy, would be invaluable. However, this is not a sine qua non for all new agents; ones with similar efficacy to existing drugs but with fewer adverse effects would be very welcome.

The requirement for drugs which do not cause tolerance to develop has been mentioned above. Simple kinetics, renal excretion or a mixture of renal and hepatic elimination is desirable, because the clinician with poor clinical pharmacological knowledge will be able to use them with fewer risks; dosing will be simpler.

Plasma drug level monitoring has become a feature of the medical management of epilepsy since sensitive analytical methods were developed in the 1960s and 70s. Unfortunately, it has become abused, often being done by rote rather than to answer a specific and sensible question. Also, it is expensive. Arguably, we could largely dispense with drug level monitoring apart, perhaps, from phenytoin whose kinetics are a particular problem (see above, and the chapter by Perucca and Pisani). Ideally, the new drug should be simple to use and dose, and also have a therapeutic ratio which is wide enough to make monitoring plasma levels unnecessary.

The use of some of the conventional antiepileptic drugs is hampered by the lack of a parenteral formulation. If a patient is unable to take carbamazepine orally because of a surgical operation or status epilepticus, the drug has to be changed or the oral formulation given by nasogastric tube. This is because an intravenous preparation is difficult to produce owing to the low water solubility of the compound. It would be much better if new drugs were adequately water soluble and the Pharmaceutical Company marketing it should be encouraged to develop a parenteral formulation at an early stage. This might also encourage early clinical assessment by single dose administration in patients with photosensitivity or interictal spikes.

The kinetics and/or mode of action of the drug should permit once or twice daily dosing. A midday dose is most inconvenient for patients and encourages non-compliance. A plasma elimination half-life of 24–48 hours is therefore the goal, unless the duration of action is prolonged by a hit-and-run type of action (eg vigabatrin and probably valproate). In my experience, most patients are quite happy with twice daily administration and once daily dosing conveys a benefit for only the minority.

Sustained release formulations have become the fashion in order to reduce the frequency of dosing. Sometimes, however, they are simply line extensions when patents expire. They may have

94

bioavailability problems (eg carbamazepine), be unnecessary because the pharmacodynamic half-life is not as short as the pharmacokinetic half-life (eg valproate), or are simply more expensive. Ideally, new drugs should have a duration of action sufficiently long to make sustained release formulations redundant.

Finally, new drugs should be inexpensive. This, of course, is rarely so because the pharmaceutical company that has developed the drug will have invested £50–100M in doing so and will wish to recoup this money in ample profits while the drug is still in patent. At the time of its introduction in the UK, vigabatrin cost roughly six times as much as carbamazepine or valproate (about £1000/year/patient) and the balance between cost and benefit became a real issue.

2. Adverse effects

In a perfect world a new antiepileptic agent would have no side effects. No such drug is known in any field of medicine. Realistically, therefore, we should look for specific features which make it superior to existing drugs. The most important are listed in Table 3.

Most of these points have been discussed above and will not be dealt with further. However, the problem of allergic reactions deserves mention because it is a common problem in the management of epilepsy. Carbamazepine has been claimed by many authors to be the least toxic of cur-

TABLE 3

ADVERSE EFFECTS – WHAT WE DO WANT

Not enzyme inducer
Not enzyme inhibitor
Not metabolized or bound to plasma proteins
Not CNS depressant
No allergic reactions
No idiosyncratic adverse effects
No teratogenicity

rently available antiepileptic drugs, yet when the withdrawal rates in controlled clinical trials are examined, carbamazepine often fares badly in comparison with other drugs. This is mainly because of skin rashes – 10–15 percent of patients experience them in monotherapy studies, and 3–5 percent in add-on studies. These are usually forgotten by clinical researchers who study adverse reactions in a chronic setting because patients who develop rashes do not proceed to chronic therapy with the offending agent. Allergic reactions to lamotrigine may be as prominent and may therefore be an important limitation of the drug.

Teratogenicity is also an important issue, particularly as all of the conventional main line agents seem to have this property. Even carbamazepine, which was thought to be relatively free of teratogenic effects, has been linked with craniofacial and other deformities recently, more than 25 years after its introduction. Valproate's ability to produce spina bifida has also come to light relatively recently, more than fifteen years after it was first used.

In general, animal testing usually identifies a teratogenic potential in a new drug. It is a requirement of all Western drug regulatory bodies that such tests are performed to their satisfaction before a new drug is considered for a clinical trials certificate. We would therefore expect new antiepileptic drugs to be free of teratogenic actions. Nevertheless, early clinical trials are usually carried out in patients who will not become pregnant during the study. In some ways this is unfortunate because it merely delays our acquisition of knowledge about the safety of the compound, as well as denying women in their childbearing years its possible benefit.

In some ways it is surprising that most of the available antiepileptic drugs, which have widely differing chemical structures, are teratogenic. Is it possible that we are underestimating the contribution of epilepsy itself to the occurrence of congenital malformations? It would be a pity if new antiepileptic drugs were accused, unfairly, of this potential when in fact, genetic and other related factors are responsible.

The new drugs

The remainder of this volume deals in turn with each of the new antiepileptic drugs under active clinical evaluation. By the time that it is published the list will be out of date because the field is moving fast. With new and exciting developments in our knowledge about glutamate and GABA transmission this pace is likely to continue for several years. The drop out rate is inevitably high, usually because of lack of efficacy in man or because of unacceptable interactions or adverse effects. There has been concern about the microvacuolation and intramyelinic oedema produced by vigabatrin, a property which appears to be common to GABA transaminase inhibitors in general, but accumulating evidence from human neuropathological studies appears to be reassuring. Although the abnormality is found in at least

three species of laboratory animal (the rat, mouse and dog), no similar lesion has so far been shown in man.

Recent concern has also been expressed (by the FDA in the USA) about the possibility that glutamate receptor antagonists might cause neuronal vacuolation, as shown with MK801. Compounds of this nature are on hold in the USA until regulatory officials have been reassured on this count.

Table 4 lists the new drugs which are undergoing clinical testing, together with their mode of action where known.

No attempt will be made here to describe each of the drugs in detail because the remaining chapters do this. However, in Table 5, I have attempted to highlight some of the possible good and bad points about the four compounds which are most advanced in clinical studies.

It must be appreciated, though, that full evaluation of a new drug takes many years, particularly with respect to the spectrum of adverse reactions because large numbers of patients need to be treated before rare adverse effects become recognised.

TABLE 4

NEW ANTIEPILEPTIC DRUGS UNDERGOING CLINICAL TESTING

Drug	Mode of action
Eterobarbital (Ciba-Geigy)	Increased GABA inhibition (?)
Felbamate (Wallace)	Not known
Gabapentin (Parke-Davis)	Not known
Lamotrigine (Wellcome)	Inhibition of glutamate release
Loreclazole (Janssen)	Not known
Losigamone (Schwabe)	Enhancement of GABA-mediated inhibition (?)
Oxcarbazepine (Ciba-Geigy)	Not known
Ralitoline (Godecke)	Inactivation of voltage dependent sodium channels (?)
Remacemide (Fisons)	Non-competitive antagonism at NMDA receptors
Stiripentol (Biocodex)	Inhibition of GABA reuptake or metabolism (?)
Tiagabine (Abbotts)	Inhibition of GABA reuptake
Topiramate (McNeil)	Not known
Vigabatrin (Marion Merrell Dow)	Inhibition of GABA-transaminase activity
Zonisamide (Dainippon)	Not known

TABLE 5

GOOD AND BAD POINTS ABOUT FOUR OF THE NEW DRUGS

Drug	Good points	Bad points
Vigabatrin	Known mode of action Simple kinetics No interactions (except for minor interaction with phenytoin) No plasma level monitoring Once daily dosing	Neurotoxicity (but probably not in man) Psychoses can be provoked Expensive (in UK)
Lamotrigine	Known mode of action Once or twice daily dosing Wide spectrum of activity?	Rashes common Interactions with other antiepileptic drugs
Oxcarbazepine	(compared with carbamazepine) Fewer rashes Fewer adverse effects? Less inducing	Mode of action not known Prodrug Half potency of carbamazepine Hyponatraemia common
Gabapentin	Simple kinetics Few adverse effects?	Mode of action not known Low efficacy? Short half life

The New Antiepileptic Drugs

New Antiepileptic Drugs (Epilepsy Res. Suppl. 3)
Editors: F. Pisani, E. Perucca, G. Avanzini, A. Richens
© 1991 Elsevier Science Publishers B.V. (Biomedical Division)

CHAPTER 11

Eterobarb

Kevin D. Wolter

Ciba-Geigy Pharma Division, Summit, NJ, USA

Introduction

Eterobarb (N,N'-dimethyoxymethy-phenobarbitone, Antilon®) was discovered in the late 1960's by a screening program at Kendall Co. that was, interestingly enough, designed to identify more potent sedative agents. Eterobarb was found to be the one phenobarbitone derivative virtually devoid of sedative-hypnotic properties in rodents, but with anticonvulsant potency similar to phenytoin and phenobarbitone. Clinical studies have since shown that eterobarb is a safe and effective antiepileptic agent with a more favourable side effects profile than currently available agents.

Chemistry

Eterobarb ($C_{16}H_{20}N_2O_5$, MW = 320.16) is a white, odourless, crystalline powder formed by the reaction of phenobarbitone and two mole equivalents of formaldehyde in the presence of an acid catalyst. One synthetic route involves isolation of the intermediate, N,N'-bis-(chloromethyl)-phenobarbitone, which is reacted with sodium methoxide to obtain the finished product, eterobarb[1-3], the structure of which incorporates one -CH_2-O-CH_3 on each of the ring nitrogens of phenobarbitone.

Eterobarb is soluble in ethanol and is 125 times more lipid soluble than phenobarbitone. Eterobarb can be assayed in biological fluids by gas chromatography using flame ionization detection[4], although HPLC methods have been developed more recently[5].

Pharmacological properties

Experiments in rodents have shown eterobarb to have similar anticonvulsant potency to other agents, in particular phenobarbitone, but eterobarb has a remarkable lack of hypnotic effects in rodents. The MES results for eterobarb indicate potency in the range of phenobarbitone. The MES ED_{50} in rats has been determined as 6.6 (4.2–10.4) mg/kg[7]. The ability of a compound to suppress MES in rodents is correlated with its ability to suppress grand mal seizures in humans[8], and this correlation has been confirmed by clinical studies of eterobarb. The durations of anticonvulsant activity were determined in mice (MES): while eterobarb (0.1 mmol/kg p.o.) was still active at 8 hours, phenobarbitone (0.1 mmol/kg p.o.) had substantially lost its activity at this time[7]. In a separate experiment, eterobarb was administered to mice at a dose of 13 mg/kg (MES ED_{50}) for four consecutive days. After 4 days, MES ED_{50} showed no change, suggesting that repeated administration did not produce either drug accumulation or tolerance[7].

100

In vivo, p.o. eterobarb is rapidly metabolised in the liver by a first pass effect to form not only phenobarbitone, but also the N-monomethoxy-methylphenobarbitone metabolite (MMMP), which structurally represents the metabolic intermediate between eterobarb and phenobarbitone[9–11]. MMMP accumulates in the brain of rats under chronic administration[12] and has been found to be an effective anticonvulsant in the MES test in the rat, although less so than eterobarb itself[11]. Eterobarb has not been detected in the serum of animals or humans after oral administration, presumably due to the rapid first pass metabolism in the liver. Thus, the critical difference between oral administration of eterobarb and that of phenobarbitone is the presence of MMMP after eterobarb. MMMP does not appear to substantially affect the anticonvulsant efficacy of eterobarb compared to phenobarbitone, but has been purported to account for the dramatic difference between these two drugs with respect to sedative-hypnotic activity.

Freer[13] characterized the development of functional tolerance to eterobarb in the rat using procedures developed by Okomoto[14–17]. Freer administered 'chronically equivalent' doses of the drug in order to maintain a reproducible degree of CNS depression throughout the dosing period. The HD_{50} for loss of righting reflex (LRR) after chronic dosing was 102 (90–116) mg/kg for phenobarbitone and 700 (569–861) mg/kg for eterobarb. There is clearly a very substantial difference in the pharmacodynamics of the two drugs, and additional oral and intravenous experiments by Freer using SKF 525A to prevent the formation of MMMP in the liver demonstrated that MMMP is responsible for the observed effects. The work of Macdonald and coworkers[18,19] on synaptic responses to GABA and glutamate in the presence of barbiturates suggests that the hypnotic effects of a barbiturate can be decoupled from the anticonvulsant effects, as MMMP appears to do to phenobarbitone.

Clinical pharmacokinetics

Intravenous eterobarb shows a bi-exponential disappearance of the drug from patients' serum with $t_{1/2}$ values of 10 and 100 minutes[20]. After oral dosing[21] eterobarb is not detectable in the serum at any time. MMMP and phenobarbitone are both detectable after oral dosing, with phenobarbitone showing a biphasic peak probably due to the formation of phenobarbitone (sequential dealkylation of eterobarb) by two pathways: directly from eterobarb (first peak), and also from MMMP (second peak). The serum half-life of MMMP is 3.3 hours, and that of phenobarbitone 5–7 days[21].

Activity and tolerability in epileptic patients and normal subjects

Eterobarb has been evaluated for efficacy in

TABLE 1

ANTICONVULSANT AND HYPNOTIC ACTIVITIES IN MICE: ETEROBARB VS PHENOBARBITONE[3,6]

Parameter	Eterobarb	Phenobarbitone
HD_{50} (mg/kg)	Not obtainable	100 (72–138)
MES ED_{50} (mg/kg)	13.5 (8–22.7)	20 (13.8–29)
MET ED_{50} (mg/kg)	57 (33–98)	25 (18.5–34)
scPTZ ED_{50} (mg/kg)	47 (29–75)	10 (7–14)

131 completed U.S. patients participating in either randomized open-label or randomized double-blind crossover protocols using phenobarbitone as a positive control[22-24]. Patients were either refractory to available treatments or unable to tolerate their current barbiturate anticonvulsant due to neurotoxicity. Patients were about evenly divided between partial complex seizures, with or without secondary generalization, and generalized tonic-clonic seizures. Eterobarb was only rarely given as monotherapy. These studies show that patients on eterobarb (i) were able to tolerate higher serum phenobarbitone levels (mean serum level = 40.6 µg/ml) without neurotoxicity compared to when they were taking phenobarbitone (mean serum level = 30.9 µg/ml), and (ii) during treatment with eterobarb overall seizure activity was significantly lower than during treatment with phenobarbitone. In the open-label studies 62 of 83 patients (75%) were improved (seizure rate) on eterobarb (p<.01 vs. phenobarbitone), while 32 of 48 (67%) were similarly improved in the double-blind studies (p<.05). Animal data does not suggest that eterobarb is a more potent anticonvulsant than phenobarbitone, but apparently the decreased neurotoxicity (sedative side effects) experienced with eterobarb allows the clinician to titrate the dose to higher, more effective, serum drug levels. The mean percent change in seizure rate seen during eterobarb treatment was a decrease of 42.9% in the double blind trials (54% in open-label trials) compared to treatment with phenobarbitone[7]. A randomized, double-blind study by Smith[25] looked at the neurotoxic effects of subchronic dosing of eterobarb in comparison to phenobarbitone in 40 normal male volunteers. His results clearly demonstrate that a given degree of neurotoxicity (sedative-hypnotic effects) will occur at markedly lower serum phenobarbitone levels in the subject given phenobarbitone as compared to the subject given eterobarb, who can tolerate much higher and more effective levels. Hyperactivity is commonly encountered among children taking phenobarbitone as antiepileptic therapy. Several cross-over studies[7,26,27], including a recent study in Italy, demonstrated not only that in children eterobarb decreases seizure rates, as in adults, but in addition, that those children suffering from hyperactivity while on phenobarbitone most often show either complete elimination of this symptom or a marked attenuation such that school performance improves.

Three cases of status epilepticus have been reported during the crossover period in the clinical trials of eterobarb; otherwise no serious or potentially serious adverse effects associated with the use of eterobarb have been reported during almost 20 years of investigational use. Sedation is the most frequently reported adverse effect. A small number of patients have been on eterobarb for over 15 years each with good results[7].

Drug interactions

As with other barbiturates, treatment with eterobarb may involve induction of the P-450 system in the liver. Concomitant use of valproate during phenobarbitone therapy may cause a marked increase in the serum levels of phenobarbitone, and this risk also applies to therapy with eterobarb, although the interaction has never been observed. Concomitant therapy with commonly used anticonvulsants has been encountered during the clinical trials with eterobarb, but not studied in detail. No significant interactions have been identified.

Conclusions

Eterobarb appears to be a promising alternative to phenobarbitone anticonvulsant therapy: a new drug with excellent efficacy and markedly attenuated neurotoxic effects. Eterobarb may be useful not only in the treatment of generalized tonic-clonic and partial motor seizures in adults, but also in children whose school performance may be impaired by the hyperactivity (attention deficit disorder) that is often associated with treatment with other barbiturates.

References

1 Samour, C., Vida, J. (1971) J. Med. Chem. 14: 187–189.
2 Vida, J., Wilber, W. (1971) J. Med..Chem. 14: 190–193.
3 Vida, J., Gerry, E. (1977) In: Medicinal Chemistry. Academic Press, New York, 15: 157–193.
4 Greely, R. (1974) Clin. Chem. 20: 192–194.
5 Fulton, B.F., personal communication: MacroChem Corp, Billerica, MA.
6 Reinhard, J.F., Reinhard, J.F. Jr. (1977) In: Medicinal Chemistry. Academic Press, New York, 15: 57–111.
7 Data on file: MacroChem Corp, Billerica, MA.
8 Woodbury, D.A. (1972) In: Purpura, D.P. et al. (eds.), Experimental Models of Epilepsy. Raven Press, New York, pp. 557–583.
9 Gallagher, B., Baumel, I., DiMicco, J., Vida, J. (1973) Fed. Proc. 32: 684.
10 Rapport, R., Kupferberg, H. (1973) J Med Chem 16: 599–602.
11 Baumel, I., Gallagher, B., DiMicco, J., Dionne, R. (1976) J. Pharmacol. Exp. Ther. 196: 180–187.
12 Gallagher, B.B., Freer, L.S. (1985) Handbook of Experimental Pharmacology, Springer-Verlag, Berlin, 74: 421–447.
13 Freer, L.S. (1978) Characterization of the functional tolerance development to N,N'-dimethoxymethyl phenobarbital. Unpublished doctoral thesis. Dept. of Pharmacy, Georgetown University, Washington, DC.
14 Rosenberg, H., Okamoto, M. (1974) Drug Addiction 3: 89–103.
15 Okamoto, M., Rosenberg, H., Boisse, N. (1975) J. Pharmacol. Exp. Ther. 192: 555–569.
16 Okamoto, M., Boisse, N., Rosenberg, H., Rosen, R. (1978) J. Pharmacol. Exp. Ther. 207: 906–915.
17 Boisse, N., Okamoto, M. (1978) J. Pharmacol. Exp. Ther. 204: 497–506.
18 Macdonald, R.L., Barker, J.L. (1979) Neurology 29: 432–447.
19 Schulz, D.W., MacDonald, R.L. (1981) Brain Res. 209: 177–188.
20 Matsumoto, H., Gallagher, B.B. (1976) In: Janz, D. (ed.), Epileptology. George Thieme, Stuttgart, pp. 122–129.
21 Jenden, D.J., Cho, A.K. (1978) Study of the Metabolism of Dimethoxymethyl Phenobarbitone, Final Report: Contract No. NC1-NS-4-2330, National Institutes of Health.
22 Gallagher, B., Baumel, I., Woodbury, S., DiMicco, J. (1975) Neurology 25: 399–404.
23 Gallagher, B.B., Woodbury, S. (1976) In: Janz, D. (ed.), Epileptology. George Thieme, Stuttgart, pp. 117–122.
24 Mattson, R., Williamson, P., Hanahan, E. (1976) Neurology 26: 1014–1017.
25 Smith, D.B., Goldstein, S.G., Roomet, A. (1976) Epilepsia, 27: 149–155.
26 Smith, D. (1977) In: Meinardi, H., Rowan, A.J. (eds.), Advances in Epileptology. Swets and Zeitlinger, Amsterdam, pp. 318–321.
27 Dalla Bernardina, B., Fontana, E., Zullini, E., Caraballo, R. (1989) Eighteenth International Epilepsy Congress, New Dehli, India.

New Antiepileptic Drugs (Epilepsy Res. Suppl. 3)
Editors: F. Pisani, E. Perucca, G. Avanzini, A. Richens
© 1991 Elsevier Science Publishers B.V. (Biomedical Division)

CHAPTER 12

Felbamate

R. Duane Sofia, Lynn Kramer, James L. Perhach, Alberto Rosenberg

Preclinical and Clinical Research and Development, Wallace Laboratories, Cranbury, NJ, U.S.A.

Introduction

Felbamate was synthesized and submitted for anticonvulsant testing to the Epilepsy Branch of the National Institute of Neurological Disorders and Stroke (NINDS) where it was entered into the Antiepileptic Drug Development (ADD) program and assigned the code ADD 03055. The results of pharmacological and toxicological studies in laboratory animals suggest that felbamate has significant and unique antiepileptic potential with a wide margin of safety.

Chemistry

Felbamate (2-phenyl-1,3-propanediol dicarbamate), whose chemical structure is shown in Figure 1, has a molecular weight of 238.24. It is a white, odourless powder which is only sparingly soluble in water, methanol, ethanol acetone and chloroform, but freely soluble in DMSO, 1-methyl-2-pyrolidinone and N,N-dimethylformamide. The substance does not exist in enantiomeric forms.

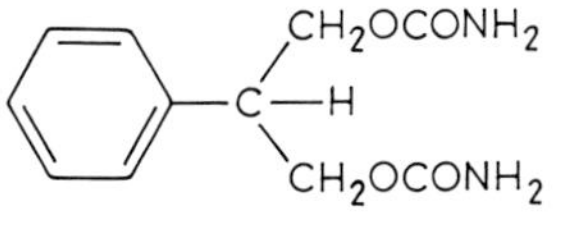

Fig. 1. Chemical structure of felbamate.

Pharmacological properties

Swinyard et al.[1,2] reported that felbamate is a drug with a unique profile of anticonvulsant action and a very low level of neurotoxicity (Table 1). Felbamate prevented seizures induced by maximal electroshock (MES), pentylenetetrazol and picrotoxin following oral or intraperitoneal administration. Felbamate inhibited bicuculline-induced seizures at higher nontoxic doses but was ineffective against strychnine-induced seizures. The effect of felbamate in chemically- and MES-induced seizure models suggests that the drug exerts its anticonvulsant activity by increasing seizure threshold and preventing seizure spread. Compared to the four prototype anticonvulsant drugs (valproate, phenytoin, phenobarbitone and carbamazepine), felbamate was the least neurotoxic. In mice (Table 2) and rats (Table 3) the oral neurotoxicity was very low; thus, felbamate has a wide margin of safety, i.e., a large protective index.

Unlike benzodiazepines, felbamate appears to have minimal potential for tolerance development. No physical dependence liability was evident following repeated administration of 300 or 600 mg/kg p.o. to rats.

Felbamate has a unique neuropharmacological profile in that it produced only weak inhibition of peripheral polysynaptic reflexes and facilitated,

TABLE 1

PROFILE OF ANTICONVULSANT ACTIVITY OF INTRAPERITONEALLY ADMINISTERED FELBAMATE AND SOME PROTOTYPE ANTIEPILEPTIC DRUGS IN MICE

Substance	TD_{50}^{*}, mg/kg	ED_{50}, mg/kg, and $[PI]^{**}$				
		MES	scMET	Bicuculline	Picrotoxin	Strychnine
Felbamate	691.99 (624.54–808.84)	40.90 (36.40–44.64) [16.92]	77.54 (61.06–114.75) [8.92]	550.03 (413.22–722.17) [1.26]	155.92 (121.84–202.40) [4.44]	no protection up to 700 mg/kg
Phenytoin	65.46 (52.49–72.11)	9.50 (8.13–10.44) [6.89]	no protection	no protection	no protection	max. protection 50% at 55–100 mg/kg
Carbamazepine	71.56 (45.91–134.79)	8.81 (5.45–14.09) [8.12]	no protection	1/2 protected 100 mg/kg	37.20 (25.32–59.69) [1.92]	78.83 (39.39–132.03) [0.91]
Phenobarbitone	69.01 (62.84–72.89)	21.78 (14.99–25.52) [3.17]	13.17 (5.87–15.93) [5.24]	37.72 (26.49–47.39) [1.83]	27.51 (20.88–34.82) [2.51]	95.30 (91.31–99.52) [0.72]
Valproate	425.84 (368.91–450.40)	271.66 (246.97–337.89) [1.57]	148.59 (122.64–177.02) [2.87]	359.95 (294.07–438.54) [1.18]	387.21 (341.37–444.38) [1.10]	292.96 (261.12–323.43) [1.45]

( ) = 95% confidence levels.
*minimal neurotoxic dose.
**protective index, TD_{50}/ED_{50}.

TABLE 2

PROFILE OF ANTICONVULSANT ACTIVITY OF ORALLY ADMINISTERED FELBAMATE AND PROTOTYPE ANTI-EPILEPTIC DRUGS IN MICE

Substance	TD_{50}^{*} (mg/kg)	MES		scMET	
		ED_{50} (mg/kg)	PI^{**}	ED_{50} (mg/kg)	PI^{**}
Felbamate	1545.0 (1299.06–1986.87)	81.1 (72.02–92.77)	19.05	548.2 (433.74–750.65)	2.82
Phenytoin	86.7 (80.39–96.09)	9.0 (7.39–10.62)	9.59	no protection	NE
Carbamazepine	217.2 (131.49–270.11)	15.4 (12.44–17.31)	14.06	48.1 (40.75–57.35)	4.52
Phenobarbitone	96.8 (79.88–115.00)	20.1 (14.78–31.58)	4.82	12.6 (7.99–19.07)	7.69
Valproate	1264.4 (800–2250)	664.8 (605.33–718.00)	1.90	388.3 (348.87–438.61)	3.26

( ) 95% confidence levels.
*Minimal neurotoxic dose.
**Protective index, TD_{50}/ED_{50}.
NE = Not evaluable.

TABLE 3

PROFILE OF ANTICONVULSANT ACTIVITY OF ORALLY ADMINISTERED FELBAMATE AND PROTOTYPE ANTI-EPILEPTIC DRUGS IN RATS

Substance	TD_{50}^{*} (mg/kg)	MES		scMET	
		ED_{50} (mg/kg)	PI^{**}	ED_{50} (mg/kg)	PI^{**}
Felbamate	>3000	47.8 (40.99–57.34)	>63	238.1 (132.07–549.34)	>12
Phenytoin	>3000	29.8 (21.92–38.91)	>100	no protection	NE
Carbamazepine	813.1 (488.75–1233.87)	8.5 (3.39–10.53)	95.65	no protection	NE
Phenobarbitone	61.09 (43.72–95.85)	9.1 (7.58–11.86)	6.68	11.5 (7.74–15.00)	5.29
Valproate	280.23 (191.32–352.76)	489.5 (351.14–728.37)	0.57	179.6 (146.73–210.35)	1.56

( ) 95% confidence levels.
*Minimal neurotoxic dose.
**Protective index, TD_{50}/ED_{50}.
NE = Not evaluable.

rather than inhibited, central reflexes following intravenous administration to cats. Furthermore, electroencephalographic changes following felbamate demonstrated stimulation of both the cerebral cortex and dorsal hippocampus.

The precise mechanism of action of felbamate is unknown. Although felbamate was shown to not significantly inhibit MK801 binding, it did significantly inhibit both NMDA- and quisqualate-induced seizures in mice. Receptor-binding studies indicated that felbamate had little or no effect on both benzodiazepine-receptor binding and γ-aminobutyric acid (GABA)-receptor binding in vitro. Moreover, in vitro felbamate had only minimal effect on adenosine uptake in mouse whole brain synaptosomes and carbonic anhydrase activity.

The anticonvulsant activity of felbamate in the MES test was antagonized by the α_1-adrenergic blocking drugs phenoxybenzamine and prazosin, whereas the α_2-adrenergic antagonist yohimbine had no effect. A similar pattern of results was observed for phenytoin.

Toxicology

Acute toxicity studies clearly demonstrate that felbamate has very low toxicity (LD_{50} > 5000 mg/kg p.o. in mice and rats). In subchronic toxicity studies (up to three months) rats and dogs were given doses up to 1000 mg/kg/day. The only unusual effect observed was crystalluria in the rat and CNS signs (sedation, ataxia and prostration) in the dog. In chronic studies (up to one year) the rat displayed an expected morphological liver change indicative of enzyme induction following an oral dose of 600 mg/kg/day. In both the rat and dog body weight gain was reduced during chronic toxicity studies. Felbamate was found to be non-oncogenic in the 104-week rat (up to 600 mg/kg/day) and the 92-week mouse (up to 1200 mg/kg/day) carcinogenicity studies. No evidence of mutagenic activity was noted in both in vitro and in vivo tests. Finally, no teratogenic or behavioural effects were observed in reproductive and teratology studies carried out in rats and rabbits. These latter findings are significant since it has been confirmed using ^{14}C-felbamate that placental transfer of the drug and its uniform distribution occurs among fetal tissues including blood.

Pharmacokinetics

Pharmacokinetics in animals

Orally administered felbamate was well absorbed in the rat, dog and rabbit. The half-life of elimination was dose-dependent, ranging from 2 to 13.4 hours. Nonlinear pharmacokinetic behaviour was demonstrated at doses higher than 400 mg/kg in the rat and 100 mg/kg in the dog and rabbit. More recently it has been shown that paediatric dogs (puppies six to eight weeks old) eliminate felbamate two to three times faster than adult dogs (one year old). Similarly, the bioavailability after a single dose as measured by area-under-the-curve was also less in paediatric versus mature dogs. Furthermore, compared to a single dose, the bioavailability of felbamate was significantly reduced in both the adult and paediatric dogs after repeated dosing for ten days, indicative of possible liver enzyme induction. Felbamate has been shown to significantly increase liver cytochrome P-450 activity.

In animals, the major route of excretion was urinary (about 66%). Faeces contains up to 23% of excretory products. Biliary excretion occurs in the rat along with enterohepatic recirculation. No accumulation of drug occurs after repeated dosing. In animals, the major metabolic pathways are hydroxylation and conjugation[3]. Unchanged felbamate was also excreted. Three metabolites have been identified, all of which are pharmacologically much less active than felbamate or inactive.

Pharmacokinetics in man

Orally administered felbamate is well absorbed in man. In both healthy subjects[3] and epileptic patients[4] the elimination half-life ranged from 14 to 22 hours following a single dose. It was slightly longer after multiple-dose administration. Peak plasma concentrations occur 1 to 3 hours post dosing. Nonlinear pharmacokinetics were observed for doses above 1600 mg/day. The major route of elimination in man is through the urine. The major metabolic pathways in man like animals are hydroxylation and conjugation.

Activity and tolerability in patients

Clinical efficacy

Open-label[4-6] and double-blind[7,8] clinical efficacy studies using felbamate as adjunctive therapy to phenytoin and/or carbamazepine in patients with partial onset seizures have been reported. Two small open-label studies noted improvement in seizure control in 75% of subjects. Wilensky et al.[4] noted a 'moderate to marked reduction in seizure frequency' in six of eight patients at felbamate doses ranging to 1600 mg/d and Sheridan[4], in doses ranging to 2400 mg/d, noted improvement in three of four patients.

An open-label trial in 15 therapy-resistant epileptic patients with partial onset seizures who were undergoing antiepileptic drug withdrawal as part of an evaluation of epilepsy surgery has been completed[6]. Three study phases were evaluated: baseline, during which antiepileptic drugs were stable; diagnostic, during which drugs were withdrawn to stimulate seizure frequency as part of the surgical evaluation; and treatment, during which the antiepileptic drug dosage given at the end of the diagnostic phase was supplemented by felbamate. Patients were reduced from an average of 1.86 drugs during the baseline period to an average of 1.06 drugs plus felbamate during the treatment period. In addition, the remaining postwithdrawal antiepileptic drugs were administered at doses less than 50% of their baseline dose. Four patients were converted to felbamate monotherapy immediately. Comparisons of seizure frequency in the baseline phase versus the treatment phase ($p<0.05$) and in the diagnostic phase versus the treatment phase ($p<0.001$) indicated greater than 50% reduction in seizure frequency in both comparisons with felbamate therapy despite the previously described significant reductions in their baseline antiepileptic drugs.

A double-blind, two-center (Universities of Minnesota and Virginia), placebo-controlled, two-period-crossover design trial in patients with therapy-resistant partial onset seizures comedicated with phenytoin and carbamazepine has been reported[7]. During the twenty-two week trial, patients received adjunctive felbamate or placebo on a t.i.d. dosing schedule to an average felbamate dosage of about 2500 mg/day. Fifty-six subjects were randomized and completed the study. Analysis was performed with percent seizure reduction ($p<0.018$), seizure frequency reduction ($p<0.046$) representing the primary variables of antiepileptic drug activity. These analyses demonstrated that felbamate produced a statistically significant reduction in seizures compared to placebo in all variables.

The second double-blind study was a placebo-controlled, double-blind, three-period-crossover trial conducted at the NIH clinical center[8]. Patients with therapy-resistant partial seizures were converted to carbamazepine monotherapy. This was followed by a three-week baseline period during which serum carbamazepine levels were maintained and seizure rates were monitored. Patients were than randomized to one of four sequences of adjunctive felbamate (F) and placebo (P) treatments (FPF, FPP, PFP, PFF). The treatments were administered over the course of three pairs of alternating two-week transition and observation periods. During the observation periods the dose of felbamate or placebo was constant. Patients received a maximum felbamate dosage of 3000 mg/day. Twenty-eight adult patients with therapy-resistant partial seizures completed the study. Fifteen of these patients identified felbamate treatment periods when asked their preference of treatment under double-blind conditions. Final analysis is pending.

Patients with uncontrolled partial onset seizures are currently being enrolled into double-blind monotherapy studies. In addition, children with Lennox-Gastaut syndrome are being evaluated in double-blind protocols.

Clinical tolerability

Adverse drug experiences are seen most frequently in polypharmacy regimens. In the two completed double-blind trials[7,8] the most frequently reported adverse experiences have been central nervous system-related (diplopia, dizziness, blurred vision, headache and ataxia) and gastrointestinally-related (nausea, vomiting and anorexia). Both studies have noted these more frequently in the felbamate treatment group, although all were felt to be clinically tolerable. Interestingly, Leppik et al.[7] found that there was not a statistical difference in these adverse experiences in comparison between felbamate treatment periods and placebo treatment periods. Theodore et al.[8] noted statistical significance in several of these categories but due to the short treatment periods, the clinical import of this result is difficult to decipher. Clinical experience has indicated a paucity of adverse experiences when comedication is reduced or monotherapy achieved. Clinical chemistry and haematology laboratory evaluation has revealed no clinically significant nonreversible abnormal result in double-blind studies.

No deaths during felbamate therapy have occurred. Except for reversible allergic rash no definite felbamate related adverse experience has been reported. In patients receiving felbamate monotherapy no significant adverse experience has been reported.

Drug interactions

Clinical trials in epileptic subjects comedicated with phenytoin and/or carbamazepine have demonstrated interactions when felbamate was added. Phenytoin plasma levels rose, in some patients 20% above baseline[4,7,9-12]. Carbamazepine plasma or serum levels have demonstrated decreases[4,7,9-12]; in one study the decreases averaged 24%[8]. The mechanisms involved in these interactions are not known.

References

1 Swinyard, E.A., Sofia, R.D., Kupferberg, H.J. (1986) Epilepsia 27: 27.
2 Swinyard, E.A., Woodhead, J.H., Franklin, M.R., Sofia, R.D., Kupferberg, H.J. (1987) Epilepsia 28: 295.
3 Wallace Laboratories, Felbamate Investigational Drug Brochure, October 1989. Carter-Wallace, Inc., Cranbury, New Jersey.
4 Wilensky, A.J., Friel, P., Ojemann, L.M., Kupferberg, H.K., Levy, R.H. (1985) Epilepsia 26: 602.
5 Sheridan, P.H., Ashworth, M., Milne, K., White, G., Santilli, N., Lothman, E.W., Dreifuss, F.E., Jacobs, M.P., Martinez, P., Leppik, I.E. (1986) Epilepsia 27: 649.
6 Leppik, I.E., Kramer, L.D., Bourgeois, R., Graves, N., Cambell, J., Cruz-Rodriquez, R. (1990) Neurology (in press).
7 Leppik, I.E., Dreifuss, F.E., Pledger, G.W., Graves, N.M., Santilli, N., Drury, I., Tsay, J.Y., Jacobs, M.P., Bertram, E., Cereghino, J.J., Cooper, B., Sheridan, P., Ashworth, M. (1989) Epilepsia 30: 661.
8 Theodore, W.H., Raubertas, R., Porter, R.J., Nice, F., Devinsky, O., Reeves, P., Bromfield, E., Ito, B., Balish, M. (1989) Epilepsia 30: 661.
9 Holmes, G.B., Graves, N.M., Leppik, I.E., Fuerst, R.H. (1987) Epilepsia 28: 578.
10 Fuerst, R.H., Graves, N.M., Leppik, I.E., Brundage, R.C., Holmes, G.B., Remmel, R. (1988) Epilepsia 29: 488.
11 Graves, N.M., Holmes, G.B., Fuerst, R.H., Leppik, I.E. (1989) Epilepsia 30: 225.
12 Graves, N.M., Ludden, T.M., Holmes, G.B., Fuerst, R.H., Leppik, I.E. (1989) Pharmacotherapy 9: 372.

New Antiepileptic Drugs (Epilepsy Res. Suppl. 3)
Editors: F. Pisani, E. Perucca, G. Avanzini, A. Richens
© 1991 Elsevier Science Publishers B.V. (Biomedical Division)

109

CHAPTER 13

Gabapentin

Marian Foot[1], Jan Wallace[2]

[1]Clinical Research Division, Warner Lambert Co., Eastleigh, Hants, UK; [2]Clinical Research Division, Parke-Davis, Ann Arbor, MI, USA

Introduction

Gabapentin is a chemically novel compound related in structure to the neurotransmitter gamma-aminobutyric acid (GABA). Increasing evidence has emerged that dysfunctions in GABAergic neurotransmission may contribute to the aetiology of epilepsy and other CNS disorders. Gabapentin was synthesised, therefore, as a GABA-mimetic that could freely cross the blood-brain barrier. The studies discussed in this paper will indicate that despite its structural relationship to GABA and anticonvulsant activity, gabapentin does not appear to act pharmacologically as a GABA-mimetic. Gabapentin is being developed as a novel broad-spectrum anticonvulsant.

Chemistry

Gabapentin (CI-945), 1-(aminomethyl)cyclohexaneacetic acid (Fig. 1), has a molecular weight of 171.34, is freely soluble in water (>10% at pH 7.4); has a pK_{a1} of 3.681 and pK_{a2} of 10.704, and does not exist in enantiomeric forms[1].

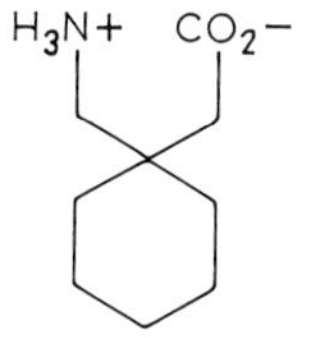

Gabapentin in the crystalline form is stable at room temperature, but a slow formation of the lactam occurs in aqueous solutions. Gabapentin may be assayed in plasma and urine by sensitive high performance liquid chromatography with pre-column labelling for ultraviolet detection[2].

Toxicology

Gabapentin was well tolerated with no deaths in acute toxicity studies at doses up to 8000 mg/kg/day in rats and up to 5000 mg/kg/day in mice. Chronic studies showed similar good tolerance with doses up to 2000 mg/kg/day in rats, up to 2000 mg/kg/day in dogs, and up to 500 mg/kg/day in monkeys. Minor increased liver enzymes and increased liver weights were seen at the highest doses in rats and dogs but not monkeys. Gabapentin was not mutagenic in bacterial and mammalian assays and was not teratogenic in three animal species.

Pharmacological properties

Animal models of epilepsy

Gabapentin has been tested in a wide variety

H_3N^+ CO_2^-

Fig. 1. Chemical structure of gabapentin.

of animal models of epilepsy[1]. Table 1 summarises the protective effect of gabapentin in various chemical and electrically induced convulsion models with mice or rats in comparison with sodium valproate.

Gabapentin was tested in male DBA/2J mice genetically susceptible to sound-induced seizures[3]. Gabapentin, given p.o. 60 minutes prior to the test, protected against wild running and/or chronic convulsions with an ED_{50} of 16 mg/kg, and from tonic extensions with an ED_{50} of 3 mg/kg. Gabapentin was also effective in protecting gerbils selectively bred to show reflex epilepsy in response to environmental change with an ED_{50} of 10 mg/kg.

Gabapentin did not show efficacy in a third genetic model using a strain of Wistar rats that show the EEG and clinical symptoms of absence seizures. Gabapentin dose-dependently aggravated the spike and wave bursts at doses of 25 and 100 mg/kg in these studies.

Gabapentin (1–20 mg/kg) showed only weak anticonvulsant activity in baboons with photosensitive epilepsy. No neurological side effects were observed, but seizures were facilitated at very high doses (240 mg/kg).

When excitatory amino acids were used as convulsants, gabapentin (30 to 420 mg/kg, i.p.) given 90 minutes prior to the convulsants, markedly prolonged the latency of onset of clonic-tonic convulsions and death in the N-methyl-D-aspartic acid (NMDA) model, but did not affect either time of onset or severity of convulsions in the kainic acid or quisqualate models[4].

Gabapentin at doses of 100 to 1000 mg/kg i.p. did not show marked anticonvulsant activity in the kindled rat model for partial seizures.

Electrophysiology

The electrophysiological effects of gabapentin on cultured spinal cord neurones in vitro were examined[5]. Gabapentin did not change postsynaptic GABA or glutamate responses, did not depress spontaneous neuronal activity, and did not block high frequency sustained repetitive action potentials at concentrations up to 30 μg/ml. This contrasts with phenytoin, carbamazepine and valproate, which may interact with voltage-sensitive sodium channels and all block sustained firing at low concentrations.

Gabapentin (3 mg/kg, i.p.) did decrease GABAergic inhibition in an experiment where paired pulse orthodromic stimulation of rat hippocampal pyramidal cells is used to evaluate GABAergic

TABLE 1

ANTICONVULSANT ACTIVITY OF GABAPENTIN ED_{50} VALUES IN mg/kg, p.o. (95% CONFIDENCE INTERVAL)

Convulsant/Test	Model	Gabapentin[*]	Valproate[†]
Semicarbazide	Inh. GABA synth.	5 (3–11)	76 (52–114)
Isoniazid	Inh. GABA synth.	20 (9–31)	325 (277–372)
3-Mercaptoproprionate	Inh. GABA synth.	31 (15–75)	71 (49–97)
Bicuculline	GABA-recep.antag.	32 (13–53)	Not tested
Picrotoxin	Cl- channel block	57 (36–89)	131 (99–163)
Strychnine	Glycine antagonist	34 (1–69)	252 (167–308)
MES	Tonic-clonic	9.4	236.6
PTZ maximal	Tonic-clonic	52 (30–75)	155 (130–181)
PTZ minimal	Absence	147 (62–1596)	83 (50–114)

[*]Gabapentin given 60 to 120 minutes prior to convulsant.
[†]Valproate given 30 to 60 minutes prior to convulsant.

mechanisms. The first pulse activates inhibitory interneurons, which inhibit or attenuate the response to a second pulse given within an inter-stimulus range of 200 milliseconds. The results obtained with gabapentin were comparable to those seen with baclofen and anticonvulsant doses of phenytoin.

Biochemical pharmacology

In the search for a mode of action for gabapentin and to test its direct GABA-mimetic activity, a wide range of receptor-binding and other in vitro studies were undertaken[6]. Gabapentin does not bind to $GABA_A$ or $GABA_B$ receptors, or to benzodiazepine, muscarinic, or glycine receptors, and it does not block sodium channels. No specific binding of gabapentin to rodent brain membrane fractions has been observed. Finally, gabapentin does not interfere with GABA metabolism. No increase in brain GABA levels in whole brain or in synaptosomal fractions was observed, nor any effect on GABA turnover or uptake.

Although gabapentin was developed as a structural analogue of GABA, it shows no direct GABA-mimetic action in the *in vitro* systems tested above. Despite this, it is clear that gabapentin enters the central nervous system and blocks seizures caused by a number of elicitors.

Clinical pharmacokinetics

In healthy volunteers, gabapentin is rapidly absorbed after oral administration. Maximum plasma levels occur 2 to 3 hours post-administration and the elimination half-life ranges from 5 to 7 hours[7]. Gabapentin is not bound to plasma proteins and is not metabolised[8]. It is excreted unchanged in urine with renal clearance approximately equalling total clearance (120–130 ml/min). The renal clearance and elimination half-life are not altered by increasing dose, al-though oral bioavailability is reduced at higher doses. Gabapentin may be titrated to full therapeutic doses in 2 to 3 days with good tolerance. The pharmacokinetics of gabapentin are not altered following multiple dosing[9]. The bioavailability of gabapentin is not affected by food. The pharmacokinetics of gabapentin were not altered in patients with epilepsy who were receiving phenytoin monotherapy[10].

Gabapentin levels in human brain are 80% of serum levels confirming animal tissue distribution studies[11].

Clinical efficacy

Human exposure

As of February 1st, 1990 over 2200 subjects have been exposed to gabapentin. Of these, 264 healthy volunteers have received the drug in pharmacokinetic studies and 158 patients have taken gabapentin as monotherapy for indications other than epilepsy, such as spasticity, and as migraine prophylaxis. Over 1800 patients with refractory epilepsy have received gabapentin in open or double-blind, placebo-controlled studies. The number of patients who have been treated long-term is extensive, with over 460 patients receiving gabapentin for longer than one year, 119 for longer than 2 years, 39 for longer than three years and 32 for longer than 4 years.

Double-blind crossover study

Early indications of the efficacy of gabapentin as an anticonvulsant came from a 25 patient dose-ranging, crossover study[12]. Following an eight-week baseline period and two week titration phase, patients were randomised to 300 mg, 600 mg or 900 mg/day of gabapentin as add-on therapy for eight weeks at each dosage in a three-way crossover design. Stable dosages of pre-existing antiepileptic drug therapy were maintained throughout the study. In addition to routine monthly monitoring of seizure frequency, side effects and serum antiepileptic drug levels, a psychometric test battery was also performed during baseline and each treatment phase.

The 15 male and 10 female patients had a mean age of 33 years (range 18 to 53) and a median duration of epilepsy of 18.5 years (range 6 to 40). Eighteen patients reported partial seizures (with and without secondary generalisation) and seven patients had primary generalised seizures. Four patients were excluded from the efficacy analysis, three patients due to questionable compliance or the addition of antiepileptic drugs during the study and one patient withdrew during the titration phase due to absence status.

The median frequency of all seizures was reduced from 3.3 to 2.1/week (45%) on 900 mg/day of gabapentin compared to baseline (p<0.001, Wilcoxon's signed rank test). There was a dose-related antiepileptic effect with the 900 mg dose significantly better than 600 mg (p=0.05) and 300 mg (p=0.01). Similar trends towards improvement with an increasing dose of gabapentin were seen for both partial seizures and tonic-clonic seizures, although gabapentin appeared to be more effective in the reduction of tonic-clonic seizures.

No trends were observed in any of the psychometric tests, indicating that no impairment of performance was noted with gabapentin therapy in the tests used.

Of the 25 patients who entered the study, eight (32%) patients reported one or more adverse events on 300 mg/day; 15 (44%) on 600 mg/day and 11 (44%) on 900 mg/day. The most common adverse events were drowsiness and tiredness. None led to withdrawal from the study, except for the patient who reported absence status during the titration period. This patient frequently reported such episodes and they had occurred previously when he had been challenged with new antiepileptic drugs.

Open, dose-titration study

Seventy patients with refractory partial and generalised epilepsies were recruited to an open study where following a 12-week baseline, gabapentin was added to the standard antiepileptic drugs in doses beginning with 300 mg/day[13]. The dose of gabapentin was individually titrated to a maximum of 1800 mg/day. Treatment continued for a period of at least two months at the optimal dose with standard antiepileptic drug treatment remaining constant throughout the study.

Eighteen patients were excluded from the efficacy sample: 13 due to poor documentation or protocol violations and five due to early withdrawals related to adverse events.

The median reduction in seizure frequency during treatment with gabapentin compared to baseline was 27% for all seizures, 32% for partial seizures, 36% for tonic-clonic seizures and 49% for absence seizures. Where response is defined as a 50% or greater reduction in seizure frequency, 29% of patients with all seizure types combined (n=52) were responders compared to 31% of patients reporting partial seizures (n=29), 35% of patients with tonic-clonic seizures (n=20) and 50% of patients with absence seizures (n=10).

Thirty-six of the 70 patients who received gabapentin reported a total of 60 adverse events. The most common events were fatigue (20%) and dizziness (12.9%). Five patients withdrew from treatment due to adverse events.

Double-blind, parallel group study

A multicentre, placebo-controlled, double-blind, parallel-group study was undertaken in patients with drug-resistant partial epilepsy reporting at least one partial seizure per week despite optimal therapy with one or two standard antiepileptic drugs[14]. Following a three-month baseline, patients were randomised to either gabapentin or placebo. Patients received 600 mg/day of gabapentin or matching placebo during a two-week

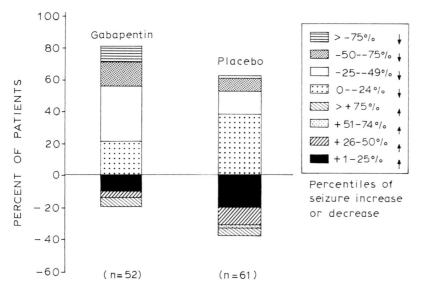

Fig. 2. Distribution of patients by percent change in partial seizure frequency.

titration period and then entered a three-month evaluation period at the full 1200 mg/day dose or placebo. Seizure frequency, adverse events, biochemistry, haematology and plasma levels of gabapentin and standard antiepileptic drugs were monitored throughout the study.

One hundred and twenty-seven patients were randomised: 61 to gabapentin and 66 to placebo. Both groups were well-matched at baseline with respect to age, duration of epilepsy, sex and baseline seizure frequency per 28 days.

Of the 127 patients recruited, 113 (61 placebo and 53 gabapentin) were included in the efficacy sample. Fourteen were excluded due to having <1 seizure per week during baseline, < 56 days diary available during baseline or treatment or because they stopped study medication for >14 days.

The evaluation of efficacy showed that 25.0% of gabapentin patients showed a reduction of at least 50% in partial seizures compared to 9.8% of placebo patients (p=0.043, Fisher's Exact Test). Figure 2 shows the distribution of responders in

25 percentile groups for all partial seizures.

Efficacy was also determined by calculating response ratios (RR). RR = (T–B)/(TB) where B is the seizure frequency per 28 days during baseline and T is the seizure frequency per 28 days during treatment. The response ratio makes the response distribution more normal as it lies in the range −1 to +1. Percentage changes in seizure frequency range from −100% to + infinity. The mean adjusted response ratio for gabapentin (−0.192) was significantly better than with placebo (−0.060, p = 0.0056). The median percent change from baseline in partial seizure frequency was greater in the gabapentin group (−29.2%) than in the placebo group (−12.5%).

Of the 61 patients who received gabapentin and the 66 who received placebo, 38 (62%) reported adverse events on gabapentin compared to 27 (41%) on placebo. The most frequent reports on gabapentin were somnolence (14.8%), fatigue (13.1%), dizziness (6.6%) and weight increase (4.9%). In the placebo group, headache (9.1%) was the most commonly reported adverse event

114

followed by dizziness (4.5%) and somnolence (4.5%). Even the most common symptoms such as somnolence and fatigue were mostly rated as mild to moderate, did not appear to be disabling and, in the gabapentin group, mostly resolved during the double-blind phase. Eleven patients, seven on gabapentin and four on placebo, withdrew from the study due to adverse events. There were no significant trends to any haematological or biochemical parameter in either treatment group.

Drug interactions

Gabapentin was shown not to induce hepatic enzymes in a controlled study comparing the effect of gabapentin and phenytoin on antipyrine kinetics in healthy volunteers[15]. Phenytoin, a known hepatic enzyme inducer, significantly increased antipyrine clearance, and significantly decreased antipyrine area under the curve (AUC).

Considering gabapentin is not metabolised, does not induce hepatic enzymes and is not protein-bound, pharmacokinetic interactions with other drugs are unlikely. To date, no interactions have been observed between gabapentin and the standard antiepileptic drugs (carbamazepine, sodium valproate, phenytoin and phenobarbitone) in clinical studies[12,13,14].

A formal interaction study of gabapentin with phenytoin in patients with epilepsy demonstrated that coadministered gabapentin does not influence plasma levels of phenytoin and vice versa[10].

Note added in proof
Toxicology data from 2-year bioassay studies conducted in rats and mice became available in September 1990 and showed an increase in acinar cell carcinomas of the pancreas in male rats. The tumours were not seen in female rats or mice of either sex. The doses of gabapentin used were up to 2000 mg/kg daily for 2 years, giving plasma concentrations of up to 85 mg/l (approx. 6 times higher than therapeutic doses given in man). In view of these findings the Company decided to

cease enrolment of patients into clinical studies pending further evaluation.

References

1 Bartoszyk, G.D., Meyerson, N., Reimann, W., Satzinger, G., von Hodenberg, A. (1986) In: Meldrum, B.M., Porter, R.J. (eds.), Current Problems in Epilepsy 4, New Anticonvulsant Drugs, John Libbey, London, pp. 147–163.
2 Hengy, H., Kolle, E. (1985) J. Chromatogr. 341: 473.
3 Bartoszyk, G.D., Fritschi, E., Herrmann, M., Satzinger, G. (1983) Naunyn-Schmied. Arch. Pharmacol. 322/Suppl. R94.
4 Bartoszyk, G.D. (1983) Naunyn-Schmied. Arch. Pharmacol. 324/Suppl. R24.
5 Taylor, C.P., Rock, D.M., Weinkauf, R.J., Ganong, A.H. (1988) Soc. Neurosci. Abstr. 14/2: 866.
6 Schmidt, B. (1989) In: Levy, R., Dreifuss, F.E., Mattson, R.H., Meldrum, B.S., Penry, J.K. (eds.), Antiepileptic Drugs, 3rd Edition. Raven Press, New York, pp 925–935.
7 Vollmer, K., Anhut, H., Thomann, P., Wagner, F., Jaehnchen, D. (1989) Adv. Epileptology 17: 209.
8 Vollmer, K., von Hodenberg, A., Kolle, E. (1986) Drug Res. 36: 830–839.
9 Tuerck, D., Vollmer, K., Bockbrader, H., Sedman, A. (1989) Eur. J. Clin. Pharmacol. 36/Suppl. A310.
10 Anhut, H., Leppik, I., Schmidt, B., Thomann, P. (1988) Naunyn-Schmied. Arch. Pharmacol. 337/Suppl. R127.
11 Ojemann, L.M., Friel, P.N., Ojemann, G.A. (1988) Epilepsia 29: 694.
12 Crawford, P., Ghadiali, E., Lane, R., Blumhardt, L., Chadwick, D. (1987) J. Neurol. Neurosurg. Psychiatry 50: 682–686.
13 Bauer, G., Bechinger, D., Castell, M. Deisenhammer, E., Egli, M., Klingler, D., Kugler, J., Ladurner, G., Lahoda, F., Mamoli, B., Schmidt, B., Spatz, R., Wieser, H.G. (1989) Adv. Epileptology 17: 219.
14 Andrews, J., Chadwick, D., Bates, D., Cartlidge, N., Boddie, G., Cleland, P., Bolel, P., Crawford, P., Shepherd, D., Yuill, G., Cumming, K., Kennedy-Young, A., Knight, R., Venables, G., Foot, M., Sauermann, W., Schmidt, B., McKenna, A., Newman, P., Saunders, M., Patterson, V. (1990) Lancet (in press).
15 Allen, E., Jawad, B., Wroe, S., Richens, A. (1987) Abstracts 17th Epilepsy International Congress, Jerusalem.

New Antiepileptic Drugs (Epilepsy Res. Suppl. 3)
Editors: F. Pisani, E. Perucca, G. Avanzini, A. Richens
© 1991 Elsevier Science Publishers B.V. (Biomedical Division)

CHAPTER 14

Lamotrigine

A.W.C. Yuen

Wellcome Research Laboratories, Beckenham, UK

Introduction

Reynolds and his colleagues[1] proposed that the antifolate effects of the major antiepileptic drugs, phenytoin, phenobarbitone and primidone, may relate to their therapeutic actions. Hence, compounds which interfere with folate metabolism might possess antiepileptic properties. Screening a series of phenyltriazines with weak antifolate activity, led to the development of lamotrigine. Further studies failed to show a correlation between the antifolate activity of compounds and their anticonvulsant activity.

Lamotrigine (Figure 1) is chemically unrelated to the currently used antiepileptic drugs and has now reached an advanced stage in its clinical development. A brief review of the chemistry, animal pharmacology, biochemical data, human pharmacokinetics and results of clinical trials will be presented.

Chemistry

Lamotrigine (BW430C) is the approved name for 3,5-diamino-6-(2,3-dichlorophenyl)-1,2,4-triazine. The empirical formula for lamotrigine is $C_9H_7Cl_2N_5$, and its molecular weight is 256.09. It is a white powder which is chemically stable, with solubility in water <1 mg/ml and in ethanol of approximately 1 mg/ml. The log P value is 1.19 at

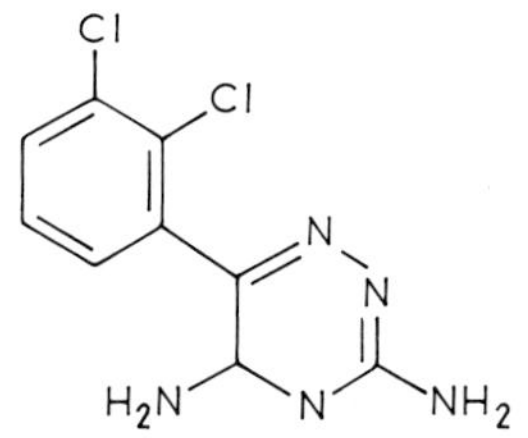

Fig. 1. Chemical structure of lamotrigine.

pH 7.6 and it has a pK_a of 5.5. Lamotrigine in body fluids can be assayed with an HPLC method[2].

Pharmacological properties

Maximal electro-shock

In acute studies lamotrigine and the known antiepileptic drugs produced a dose dependent abolition of hind limb extension (HLE) in mice[3] and rats[4]. Lamotrigine was the most potent of the drugs tested with a similar potency to phenytoin and diazepam in mice (Table 1) and to carbamazepine in rats. Phenobarbitone was moderately active in both species and valproate, as expected, was extremely weak. The duration of action of lamotrigine was long in both mice and rats as indicated both by similar ED_{50} values over the 1–8 h period post-drug and also by having only a moderate increase ($\times 3$) in the ED_{50} value at 24 h compared with the value at peak activity. All the

TABLE 1

MAXIMUM ELECTRO-SHOCK TEST IN MICE. ORAL ED_{50} VALUES (MG/KG) FOR HLE ABOLITION

Drug	ED_{50} at peak activity (Times post-drug)	ED_{50} at 24 h post-drug
Lamotrigine	2.6–3.8 (1–8 h)	10.1
Phenytoin sodium	3.5–5.2 (4–8 h)	35.9
Phenobarbitone sodium	9.1–11.4 (1–2 h)	64.7
Carbamazepine	6.9–8.5 (0.25–1 h)	352
Valproate sodium	332–461 (0.25–2 h)	Not tested
Diazepam	3.2–5.5 (0.25–2 h)	119
Troxidone	750–609 (0.25–1 h)	Not tested
Ethosuximide	Inactive	

known antiepileptic drugs were less persistent with the exception of phenytoin in the mouse: although good potency was seen up to 8 h post-drug the 24 h ED_{50} was approximately 10 × peak value.

No tolerance to the anticonvulsant action of lamotrigine was observed following administration (p.o.) for up to 28 days in both mice and rats with electro-shock applications once weekly to determine ED_{50} values.

Lamotrigine was effective by all routes tested (p.o., s.c., i.p. and i.v.) with similar potencies. In rats, for example, the respective ED_{50} values (with time intervals from drug to shock) were: 2.0 mg/kg p.o., 2.4 mg/kg s.c., 3.6 mg/kg i.p. (all at 2 h) and 3.0 mg/kg i.v. (1 h).

Chemical convulsants

Lamotrigine was tested against maximal and threshold seizures induced in mice following intravenous infusion of leptazol[3] and bolus injection of picrotoxin and bicuculline. In all the maximal seizure tests, lamotrigine, and the known drugs, with the exceptions of ethosuximide and troxidone (both examined only in the leptazol test), abolished or reduced the incidence of HLE. The ED_{50} values for HLE abolition were similar to those found in the MES test. Lamotrigine and phenytoin failed to increase clonus latencies in threshold tests, and at high doses (160 and 640 mg/kg respectively), the drugs lowered clonus latencies indicating a pro-convulsant action at these doses.

After discharge tests

Lamotrigine reduced electrically-evoked after discharge (AD) duration in a dose dependent manner with ED_{50} values in the rat and dog of 11.7 and 4.5 mg/kg iv respectively. Phenobarbitone was effective in the rat and dog whilst phenytoin was only effective in the dog[5].

Lamotrigine inhibited visually evoked AD in the rat, in a dose dependent manner with an ED_{50} value of 3.2 mg/kg[6]. Diazepam and ethosuximide were effective against visually-evoked AD but ineffective in the MES test. Conversely, phenytoin and carbamazepine were ineffective in reducing visually-evoked AD but were effective in the MES test.

Epileptic kindling

The effect of lamotrigine on the development of the electrically induced cortical kindling was tested in rats[7]. Lamotrigine produced a highly significant dose related increase in the number of nil responses, i.e. complete absence of behavioural or EEG responses to electrical stimulation. The percentage of nil responses in controls and at 3, 6, 12 and 18 mg/kg of lamotrigine were 2, 26, 42,

67 and 86 percent respectively. Although the drug failed to block kindling development, the two higher doses significantly decreased both the number of kindled responses and their duration.

Rotarod

In a test of motor incoordination in mice, lamotrigine, phenytoin and control fluid were administered p.o. under randomised blind conditions. The mice were assessed 2 h after drug administration for ataxia using a 4 point rating scale. Lamotrigine showed a 35 fold separation between the anticonvulsant ED_{50} value and the dosage to induce ataxia as compared with only an 11 fold separation for phenytoin.

Neurochemical studies

Veratrine (4 and 10 μg/ml) and potassium (55 mM) produce a marked release of glutamate and GABA with only a moderate release of aspartate from rat cerebral cortical slices. Lamotrigine inhibited veratrine-evoked release of glutamate and aspartate with an ED_{50} value of 21 μM but lamotrigine up to 300 μM had no effect on potassium-evoked amino acid release[8].

The total lack of effect of lamotrigine on potassium-evoked release and the potent effect on veratrine-evoked release at anticonvulsant brain concentrations, suggests that lamotrigine is blocking voltage sensitive sodium channels to stabilise neuronal membranes and hence inhibit neurotransmitter release, principally glutamate.

Clinical pharmacokinetics

Absorption and bioavailability

After oral administration, lamotrigine is completely absorbed with negligible first pass effect. In a radiolabel study in which 6 healthy volunteers received lamotrigine 240 mg containing 15 μCi of ^{14}C-tracer, mean ± SD total radioactivity recovery values over 168 h were 94 ± 2% in urine and only 2 ± 0.5% in faeces. In an intravenous versus oral crossover study in 8 volunteers, using a 75 mg dose, the mean ± SD absolute bioavailability (F) of the oral formulation (capsule) was 0.98 ± 0.05.

Plasma pharmacokinetics

The plasma pharmacokinetics of single oral doses of lamotrigine have been obtained from a number of studies involving healthy volunteers. Plasma lamotrigine concentrations are described adequately by a one compartment open pharmacokinetic model, with first order absorption and elimination. Lamotrigine is absorbed rapidly with average values for lag time and absorption rate constant of 0.3 h and 9 h^{-1} respectively, and Tmax generally between 2.2–3 h. Individual plasma profiles often show multiple peaks which have been attributed to such factors as salivary excretion and sequestration in the stomach[9]. Single dose studies in healthy volunteers, using lamotrigine 15, 30, 60, 120 and 240 mg have demonstrated the linear kinetics of lamotrigine, with peak plasma concentration (Cmax) and area under plasma concentration curve (AUC) shown to be directly dose proportional[2]. Typical Cmax and AUC values after 120 mg single oral dose in healthy volunteers are 1.6 μg/ml and 53 μg/ml/h respectively.

Clearance and elimination half-life of lamotrigine showed considerable intersubject differences due to variable rates of metabolic clearance, but intrasubject variation is much less marked. Typical values for clearance and elimination half life in single oral dose studies in healthy volunteers are 42 ml/min and 24 h respectively[2]. In the multiple oral dose study, ten healthy male volunteers received lamotrigine 120 mg b.i.d. on day 1 and 60 mg b.i.d. on days 2 to 7. Average plasma and saliva lamotrigine elimination half life values were 26 h and there was no evidence of a change in clearance over the one week dosing period[2].

Metabolism

Lamotrigine is extensively metabolised in man, mainly to a β-glucuronidase hydrolysable glucuronide. In single oral dose studies in healthy volunteers with sufficiently long urine collection periods, an average of 8% of the dose was recovered in urine as unchanged lamotrigine and 63% as the β-glucuronidase hydrolysable glucuronide[2]. Similar values were obtained after intravenous administration[9].

Excretion

The renal clearance of lamotrigine itself is only 0.04 ml/min/kg, accounting for less than one-tenth of the total clearance of the drug. However, the limited data available indicate that the major metabolite must have a high renal clearance, possibly approaching glomerular filtration rate, and that the rate of excretion of glucuronide in urine is limited by its rate of formation.

Activity and tolerability in patients with epilepsy

Single dose studies

In an open study[10], five patients with frequent interictal spike activity showed a reduction in the frequency of spikes following a single oral dose of lamotrigine 120 mg or 240 mg. The reduction ranged from 78 to 98%.

In another study[11] six patients with frequent interictal spikes took part in a randomised double-blind 3 way crossover study with placebo, diazepam 20 mg and lamotrigine 240 mg. Diazepam 20 mg significantly reduced spike counts compared with placebo at all post drug times from 0.5 h. Lamotrigine 240 mg significantly reduced counts at all times from 1 h. Lamotrigine's onset of action with respect to interictal spike suppression was slower than diazepam but like diazepam it was maintained for at least 4 hours.

In an open study[10], the effect of a single oral dose of lamotrigine 120 mg or 240 mg on the photoconvulsive response was studied in six photosensitive patients. All six photosensitive patients showed a reduction in the photosensitivity range and in two of the patients photosensitivity was abolished.

One week studies

In an open study[12], 20 patients received daily or twice daily lamotrigine for a week. The dose given ranged from 50 mg once a day to 225 mg twice a day. The Wilcoxon signed ranks test showed that there were significantly fewer seizures ($p < 0.01$) in the lamotrigine treatment week compared to the baseline week for both complex partial seizures and all seizure types combined. Some patients showed a marked increase in seizure frequency on stopping lamotrigine. This may be due to rebound seizures on stopping lamotrigine abruptly.

Another one week lamotrigine study[13] was conducted as a double-blind placebo controlled crossover add-on trial in patients with treatment resistant epilepsy. Ten patients were entered and the total daily dose of lamotrigine given ranged from 100 to 300 mg. Patients were taking two to four concomitant antiepileptic drugs which were maintained unchanged throughout the study. Six patients showed at least 50% seizure reduction on the lamotrigine treatment week compared with the placebo week. One had less than 50% seizure reduction, one no change and two showed an increase in seizures. While not statistically significant ($p = 0.055$), there was a clear trend for a reduction in seizures during the lamotrigine treatment week. There was a significant reduction in epileptiform activity on lamotrigine compared with placebo.

Efficacy studies

A total of 4 randomised double-blind placebo controlled crossover trials of lamotrigine added on to a regimen of standard antiepileptic drugs, in patients with refractory seizures, have been completed and analysed. The first three performed in Holland[14], Cardiff[15] and Chalfont[16], had the same design with five phases: Baseline, 8 weeks; Treatment Period 1, 12 weeks; Washout 1, 6 weeks; Treatment Period 2, 12 weeks; Washout 2, 6 weeks. The fourth study in Bordeaux[17] has shorter phases.

In Holland, 30 out-patients with refractory partial seizures completed the study. Patients continued their standard antiepileptic drugs: 2 patients were on 4 drugs, 12 were on 3 drugs, 14 patients were on 2 drugs and 2 were on 1 drug (mean number of antiepileptic drugs = 2.5). The lamotrigine dose used ranged from 50 to 400 mg total daily dose. This produced an estimated mean trough plasma lamotrigine concentration of 1.47 μg/ml. Nineteen out of 30 patients showed an improvement on lamotrigine treatment.

In the Cardiff study, 21 out-patients with refractory partial seizures completed the study. The patients continued on their standard antiepileptic drugs: 17 patients were on 2 drugs and 4 were on 1 drug (mean number of antiepileptic drugs = 1.8 per patient). The dosage strategy in this study aimed at achieving trough plasma lamotrigine concentrations of 1.5 to 2.5 μg/ml. The daily dosage of lamotrigine used in the study ranged from 75 to 400 mg. This produced a mean trough plasma lamotrigine concentration of 1.9 μg/ml for the 21 patients in the study. Eighteen out of 21 patients showed an improvement on lamotrigine treatment.

In Chalfont, 18 residential patients with refractory seizures completed the study. Patients continued on their standard antiepileptic drugs: 10 patients were on 3 drugs, 6 were on 2 drugs and 2 were on 1 drug (mean number of antiepileptic drugs = 2.4). The daily dosage of lamotrigine used ranged from 100 to 300 mg. This produced a mean trough plasma lamotrigine concentration of 2.1 μg/ml. Twelve out of 18 patients had fewer seizures during the lamotrigine treatment period compared with placebo.

In Bordeaux, 23 out-patients with refractory partial seizures completed the study. Patients continued on their standard antiepileptic drugs: 15 patients were on 2 drugs and 8 were on 1 drug (mean number of antiepileptic drugs = 1.7). The daily dosage of lamotrigine used ranged from 75 mg to 300 mg. This produced a mean plasma lamotrigine concentration of 1.47 μg/ml. Fifteen out of 23 patients showed an improvement on lamotrigine treatment.

The four trials were analysed according to the two period crossover model by Koch[18] using the non parametric Wilcoxon rank sum test. An estimate of lamotrigine's treatment effect on total seizures was obtained by calculating the percent seizure reduction given as Lamotrigine/Placebo × 100. The median percent reduction in total seizures and its 95% confidence intervals were then derived using the binomial theorem and order statistics. The results of the 4 trials are summarised in Table 2 and Figure 2.

The data from the 4 trials were pooled in a meta-analysis to provide an overview of the results and a more precise estimate of the treatment effect. Of the 92 patients, 64 (70%) had lower total seizures in the lamotrigine treatment period compared with the placebo treatment period. Twenty-five patients (27%) had a >50% reduction in seizures compared with placebo. The test for treatment effects shows a statistically significant difference in favour of lamotrigine ($p<0.001$). The median seizure reduction for total seizures was 27% with a confidence interval of 19% to 34% and a confidence coefficient of 0.941.

The efficacy data clearly show that lamotrigine has antiepileptic efficacy and is effective in reducing partial and generalised tonic-clonic seizures unsatisfactorily controlled with standard AEDs.

The adverse experiences from the 4 controlled

TABLE 2

RESULTS OF 4 CONTROLLED STUDIES ON LAMOTRIGINE

	No of patients	Total seizures	Partial seizures	Generalised tonic-clonic seizures	Median percent reduction in total seizures	95% confidence intervals	
Holland	30	p<0.02	p<0.01	NA	17	0	30
Cardiff	21	p<0.002	p<0.001	p<0.05	59	34	76
Chalfont	18	NS	NS	p<0.05[a]	18	−14	29
Bordeaux	23	p<0.05	p<0.05	NA	23	−11	52

NS: non significant; NA: not analysed, insufficient patients; a: analysis of last 8 weeks of treatment periods

trials were pooled and Table 3 shows the incidence of adverse experiences during the lamotrigine and placebo treatment periods. Although the incidence of headache and ataxia appears greater on lamotrigine treatment, statistical analysis showed that none of the adverse experiences occurred significantly more frequently on lamotrigine treatment. Analysis of adverse experiences in individual trials also did not show any significant increase in incidence on lamotrigine treatment.

Examination of other safety variables including, physical, neurological examination, ECG, vital signs, haematology and biochemistry failed to reveal any changes attributable to lamotrigine treatment.

Open studies

A large number of open add-on studies have been performed to extend the experience with the use of lamotrigine. The data from 27 open studies of similar design, including 572 patients have been pooled. Seizure counts by seizure types

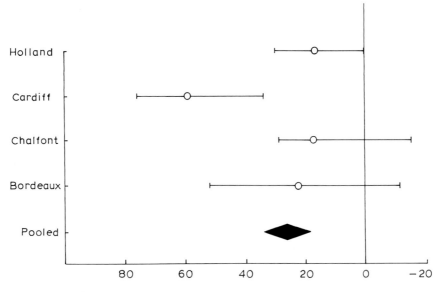

Fig. 2. Percent seizure reduction.

TABLE 3

ADVERSE EXPERIENCES

Adverse experience	Incidence on placebo	Incidence on lamotrigine	95% CI for PLO-LTG
Asthenia	13%	17%	−12% ; 4%
Diplopia	12%	15%	−13% ; 6%
Headache	7%	13%	−14% ; 1%
Somnolence	12%	13%	−9% ; 7%
Ataxia	4%	12%	−15% ; 0%
Dizziness	8%	9%	−9% ; 7%
Nausea	3%	4%	−7% ; 5%
Nervousness	2%	4%	−7% ; 3%

were recorded during these studies. Changes in seizure frequency in the first 12-week period has been compared with a retrospective 12-week baseline. A summary of the results is shown in Table 4.

Overall 30% of patients showed at least a 50% reduction in total seizures. This supports the findings in the controlled trials. For generalised tonic-clonic seizures, 52% of patients showed at least 50% seizure reduction and 22.5% of patients with tonic-clonic seizures were completely seizure-free of their tonic-clonic seizures.

There were much smaller numbers of patients with other seizure types but the limited data are encouraging for absence, atypical absence, myoclonic and atonic seizures with approximately 50% of patients showing at least 50% seizure reduction.

A proportion of patients who have not shown a reduction in seizure frequency have nevertheless benefited as their seizures were less severe and they experienced less post-ictal disturbance. A number of patients have reported a sense of well being on lamotrigine treatment even when they have not experienced improved seizure control. These effects are currently being evaluated in a controlled study.

Over 1600 patients have now been given lamotrigine in clinical trials, and from a database of 1431 patients treated with LTG mainly as add-on therapy in open studies, 123 (8%) were withdrawn with adverse experiences (Table 5). The

TABLE 4

OPEN STUDIES: PERCENTAGE OF PATIENTS WITH ≥ 50% SEIZURE REDUCTION IN THE FIRST 12 WEEKS OF LAMOTRIGINE TREATMENT

Seizure type	%	Total patients assessable
All	30	417
Partial	28	308
Tonic-Clonic	52	209
Absence	46	24
Atypical absence	64	11
Myoclonic	36	11
Atonic	50	14

most common cause for patient withdrawal was skin rash occurring in 21 patients giving a rate of approximately 2.5%. The next most common withdrawal reasons were diplopia and seizure increase, each of which occurred in less than 1% of patients.

Drug interactions

In vitro studies have shown that phenytoin, phenobarbitone and valproate at therapeutic concentrations have no significant effect on the degree of plasma protein binding of lamotrigine. Conversely plasma lamotrigine concentrations of 2.5–3 μg/ml do not alter the plasma protein binding of phenytoin, phenobarbitone, carbamazepine and valproic acid at therapeutic concentrations.

Nevertheless, concomitant treatment with other antiepileptic drugs can have a substantial effect on the metabolic clearance of lamotrigine.

In a single oral dose study of lamotrigine in patients, the mean ± SD lamotrigine elimination half-life of 4 patients on sodium valproate was 59 ± 26 h[10]. Although the number of patients was small, sodium valproate appeared to inhibit the metabolism of lamotrigine and prolonged the lamotrigine elimination half life.

On the other hand, in both single and chronic dosing studies, the elimination half-life of lamotrigine in patients taking liver enzyme-inducing antiepileptic drugs such as carbamazepine, phenytoin, phenobarbitone and primidone may be shortened to values approximately half those generally found in healthy volunteers receiving no other medication[10,12]. The combination of sodium valproate and enzyme inducing drugs appears to 'normalise' the elimination half life of lamotrigine towards values found in healthy volunteers[12].

Increases in the plasma concentrations of carbamazepine, phenytoin and valproate have been reported in individual patients when lamotrigine

TABLE 5

PATIENTS WITHDRAWN DUE TO ADVERSE EXPERIENCES

Adverse experience	No of patients
Rash	39
Diplopia	13
Increase in seizures	11
Headache	7
Nausea & vomiting	7
Ataxia	5
Drowsiness	4
Aggression	3
Depression	3
Dizziness	3
Asthenia	2
Confusion	2
Syncope	2
Adverse experience reported by one patient only	22
Total	123

Total number of patients assessed = 1431.

has been added on to their existing antiepileptic treatment regimen. However, formal analysis of single dose, one week and 12 week administration of lamotrigine has shown no significant change in the plasma concentrations of other antiepileptic drugs[10,12,13,14,15].

References

1 Reynolds, E.H., Milner, G. Matthews, D.M., Chanarin, I. (1966) Quart. J. Med. 35: 521–537.

2 Cohen, A.F., Land, G.S., Breimer, D.D., Yuen, W.C., Winton, C., Peck, A.W. (1987) Clin. Pharmacol. Ther. 42: 535–41.

3 Lamb, R.J., Leach, M.J., Miller, A.A., Wheatley, P.L. (1985) Br. J. Pharmacol. 85: 235.

4 Miller, A.A., Sawyer, D.A., Roth, B., Peck, A.W., Leach, M.J., Wheatley, P.L., Parsons, D.N., Morgan, R.J.I. (1986) in: Meldrum, B.S., Porter, R.J. (eds.), New Anticonvulsant Drugs. John Libbey, London, 165–177.

5 Miller, A.A., Wheatley, P.L. (1985) Br. J. Pharmacol. 85: 366.

6 Lamb, R.J., Miller, A.A. (1985) Br. J. Pharmacol. 86: 765.

7 Leach, M.J., Miller, A.A., O'Donnell, R.A., Webster, R.A. (1983) J. Neurochem. 41: 1492–1494.

8 Leach, M.J., Marsden, C.M., Miller, A.A. (1986) Epilepsia 27: 490–7.

9 Yuen, W.C., Peck, A.W. (1987) Epilepsia 28: 582.

10 Binnie, C.D., Van Emde Boas, W., Kasteleijn-Nolste-Trenite, D.G.A., Korte, R.A., Meijer, J.W.A., Miller, A.A., Overweg, J., Peck, A.W., Wieringen, A., Yuen, W.C. (1986) Epilepsia 27: 248–54.

11 Jawad, S., Oxley, J.R., Yuen, W.C., Richens, A. (1986) Br. J. Clin. Pharmacol. 22: 191–193.

12 Jawad, S., Yuen, W.C., Peck, A.W., Hamilton, M.J., Oxley, J.R., Richens, A. (1987) Epilepsy Res 1: 194–201.

13 Binnie, C.D., Beinteima, D.J., Debets, R.M.C., Van Emde Boas, W., Meijer, J.W.A., Meinardi, H., Peck, A.W., Westendorp, A.M., Yuen, W.C. (1987) Epilepsy Res 1: 202–208.

14 Binnie, C.D., Debets, R.M.C., Engelsman, M., Meijer, J.W.A., Meinardi, H., Overweg, J., Peck, A.W., Van Wieringen, A., Yuen, W.C. (1989). Epilepsy Res 4: 222–229.

15 Jawad, S., Richens, A., Goodwin, G., Yuen, W.C. (1989) Epilepsia 30: 356–363.

16 Sander, J.W.A.S., Patsalos, P.N., Oxley, J.R., Hamilton, M.J., Yuen, W.C. (1990) Epilepsy Res (in press).

17 Loiseau, P. (1990) in preparation.

18 Koch, G.G. (1972) Biometrics 1972; 577–584.

New Antiepileptic Drugs (Epilepsy Res. Suppl. 3)
Editors: F. Pisani, E. Perucca, G. Avanzini, A. Richens
© 1991 Elsevier Science Publishers B.V. (Biomedical Division)

125

CHAPTER 15

Loreclezole

F. De Beukelaar, L. Tritsmans

Janssen Research Foundation, Beerse, Belgium

Introduction

Loreclezole is a triazole derivative, synthetized by the Janssen Research Foundation, Beerse (Belgium). It is chemically unrelated to existing antiepileptic drugs. The extensive pharmacological studies and initial clinical trials that have been carried out on this compound are summarized in this chapter.

Chemistry

The structural formula of loreclezole [(Z)-1-[2-chloro-2-(2,4 dichlorophenyl)ethenyl]-1H-1,2,4-triazole] is shown in Figure 1. The solubility in aqueous medium is lower than 1 mg/100 ml. The log P value is 3.61 at pH 6.

Pharmacological properties

Anticonvulsant effects
The anticonvulsant potential of loreclezole was

assessed by comparing its activity with that of prototype anticonvulsants in well established animal models of epilepsy (see Table 1)[1-4]. The anticonvulsant effects of loreclezole were observed in different species (mice, rats, guinea-pigs and dogs) and different models of epilepsy (genetic audiogenic seizures, seizures elicited chemically or electrically).

The anticonvulsant profile of loreclezole is similar to that of phenobarbitone and diazepam: it antagonizes clonic and tonic convulsions, whereas phenytoin and carbamazepine only antagonize the tonic seizure component[1,4]. Loreclezole therefore can be considered a potentially broad-spectrum anticonvulsant.

The safety margin of the substance was studied in three animal species. In mice (rotarod test), the toxic/active dose ratio was close to 1, whereas it was 4 in rats and 10 in dogs (righting reflex and/or ataxia). This finding suggests that the separation between the anticonvulsant and the neurotoxic dose of loreclezole is highly dependent upon animal species[4].

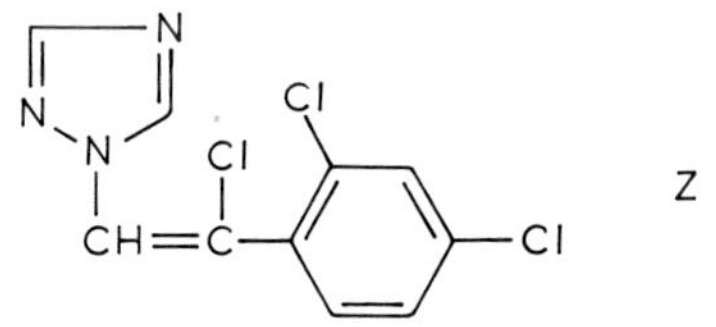

Toxicology data
No toxic effects were noted in animal toxicology studies that would preclude administration of loreclezole to man. Reproduction studies did not show primary adverse effects on the offspring. In rabbits, loreclezole was not shown to be

Fig. 1. Chemical structure of loreclezole.

TABLE 1

ED_{50} VALUES (MG/KG) RELATIVE TO PROTECTION AGAINST TONIC HINDPAW EXTENSION (THP), TONIC FOREPAW EXTENSION (TFP) AND CLONIC SEIZURES (CLON) OBTAINED WITH LORECLEZOLE GIVEN P.O. TO DIFFERENT SPECIES

Seizure test	Species	Seizure components		
		THP	TFP	CLON
Genetic model				
Audiogenic	mouse	9.2	(*)	22.3
Threshold tests				
Pentylenetetrazol (s.c.)	mouse	–	–	31.7
Bicuculline (s.c.)	mouse	–	–	15.6
Pentylenetetrazol (s.c.)	rat	–	–	17.9
Pentylenetetrazol (s.c.)	guinea-pig	–	–	5.7
Maximal seizure tests (chemical stimulation)				
Max pentylenetetrazol	mouse	24	–	–
Bicuculline (i.v.)	mouse	22.2	22.2	22.2
Max pentylenetetrazol	rat	12.2	20	20
Allylglycine	rat	9	12	25
Bicuculline (i.v.)	rat	28.4	38.5	>40
Max pentylenetetrazol	dog	0.70	3.78	1.57
Maximal seizure tests (electrical stimulation)				
Electroshock	mouse	84.0	–	–
Electroshock	guinea-pig	–	56.6	–

(*) TFP was not evaluated in this test.

teratogenic, which was in contrast to some other antiepileptic drugs. No mutagenic potential has been demonstrated.

Pharmacokinetics

Single-dose pharmacokinetic data indicate that loreclezole is rapidly absorbed after oral administration and that the apparent clearance (oral dose divided by $AUC_{0-\infty}$) of the drug is several times higher (73 l/day) in patients receiving loreclezole in addition to existing antiepileptic medication than in normal volunteers (8.5 l/day)[7].

Multiple-dose studies in patients taking other antiepileptic drugs demonstrate a terminal half-life between 9.8 and 33.8 days[6].

Activity and tolerability studies in man

Safety and tolerability in volunteers

Phase-I studies in human volunteers indicate that sedation and coordination disturbances may occur at plasma levels above 10 μg/ml[7]. Furthermore, the multiple sleep latency test (MSLT) was used to investigate whether loreclezole has sedative effects at plasma concentrations below 10 μg/ml. No statistically significant differences between loreclezole- and placebo-treated volunteers were found. No subjective changes in alertness or mood, concentration and general well-being were noted after intake of loreclezole. No other adverse events have been reported.

Single-dose efficacy studies in photosensitive patients

Photosensitivity can be used as a model in short-term studies to assess the efficacy of new antiepileptic drugs in man. As a quantitative measure, the photosensitive range (Hz) was used at which generalized paroxysmal activity on the EEG was seen. Five patients with a consistent and easily reproducible photoparoxysmal response to intermittent light stimulation participated in this single-dose experiment. A single dose of loreclezole between 100 and 150 mg caused a clear reduction in the photosensitive ranges in three patients and an actual abolition in two. In all five patients, the onset of activity coincided with the peak plasma concentration of loreclezole and the effect lasted for 24 hours.

These observations clearly indicate that loreclezole readily passes the blood-brain barrier and exerts its activity on the central nervous system. No adverse experiences were reported[8]. A comparison with other anticonvulsants in this model showed that loreclezole exerts similar effects to those of other prototypal drugs[9].

Multiple-dose efficacy studies with loreclezole

To date, open and double-blind studies with loreclezole have been performed in more than 100 patients with predominantly partial epilepsy, with or without secondary generalization.

The efficacy of loreclezole is likely to be more dependent on plasma level than dose. For this reason, the clinical trials reported here all made use of frequent plasma level monitoring to keep loreclezole concentrations within preset limits.

In a first open trial in 13 therapy-resistant patients, the efficacy of loreclezole was evaluated at a target plasma concentration of 1–2 µg/ml. The treatment period was 6 months. Seizure frequency decreased in 11 out of 13 patients, and in 4 of these the reduction was clinically significant ($\geq 50\%$).

One large controlled study has been completed so far. Sixty-two therapy-resistant epileptic patients (the majority already taking 2 or more antiepileptic drugs) were included in this double-blind trial, which evaluated the efficacy of loreclezole at a target plasma level of 1–2 µg/ml. After a 3-month baseline period, they received either loreclezole or placebo as add-on therapy to their regular antiepileptic regimen for three months. After three months, the best response in the placebo group was a decrease in seizure frequency of 42%, whereas 8 patients in the loreclezole group clearly did better (Fisher's exact test: 8/32 versus 0/30; p=0.005). The mean ($\pm$SD) plasma level of loreclezole reached after 3 months in the double-blind phase of the study was 1.69 ($\pm$0.43) µg/ml.

The patients were then allowed to enter an open follow-up study in which higher doses of loreclezole were used. Fifty-three patients were thus treated for a median duration of 352 days. Increasing loreclezole plasma concentrations clearly resulted in a gain in efficacy. The median daily seizure frequency decreased from 0.39 at baseline to 0.26 seizures/day at the end of treatment (p<0.0001, Wilcoxon MPSR test). Eventually 23 out of 53 patients (42%) could be considered responders ($\geq 50\%$ decrease). The mean loreclezole plasma concentration reached in this study was 5.7 µg/ml. These results clearly demonstrate the long-term efficacy of loreclezole in a therapy-resistant population.

No pharmacokinetic interactions with the pre-existing antiepileptic drugs were observed during these clinical trials.

Tolerability in patients

In general, loreclezole was very well tolerated and no serious adverse events were reported.

128

The safety profile of loreclezole after chronic oral administration was evaluated in 111 therapy-resistant epileptic patients. Loreclezole was always given as add-on therapy to other antiepileptic drugs. Twelve patients were treated for more than a year, 46 between 8 and 12 months, 36 between 4 and 8 months and 17 for less than 3 months. A mean ($\pm$SD) loreclezole level of 5.7 ($\pm$2.3) μg/ml (n=102) was reached at the end of treatment.

Except for an increase in γGT, which was already high before loreclezole was administered, no abnormalities were found in biochemical and haematological parameters under long-term add-on loreclezole treatment[10].

Drug interactions

As reported above, the apparent oral clearance of loreclezole appears to be increased by concomitant treatment with other antiepileptic drugs.

References

1 Wauquier, A., Fransen, J., Melis, W., Ashton, D., Gillardin, J.M., Van Clemen, G. (1987). Preclinical Research Report (PRR) R 72 06312. Janssen Research Foundation (JRF), Access no. N56239.
2 Wauquier, A., Melis, W., Van den Broeck, W.A.E. (1988). PRR R 72 o63/6. JRF, Access no. N59941.
3 Wauquier, A., Smeyers, F., Melis, W., Woestenborghs, R. (1988). PRR R 72 063/7. JRF, Access no. N59942.
4 Megens, A.A.H.P., Lenaerts, F.M., Awouters, F.H.L., Niemegeers, C.J.E. (1989). JRF, Access no. N69433.
5 Van de Velde, V., de Beukelaar, F., Van Peer, A., Van Rooy, P., Woestenborghs, R., Heykants, J., Vanden Bussche, G. (1988). JRF, Access no. N62347.
6 Van Peer, A., Van de Poel, S., Woestenborghs, R., Heykants, J., de Beukelaar, F., Smakman, J., Arends, J., Brunia, C.H.M. (1990).CRR R 72 063/17. JRF, Access no. N71455.
7 Brunia, C.H.M., Arends, J., de Beukelaar, F., Smakman, J., Tritsmans, L. (1989). CRR R 72 063/13. JRF, Access no. N71375.
8 Overweg, J., de Beukelaar, F. (1990). Epilepsy Res. (in press).
9 Binnie, C.D., Kasteleijn-Nolst Trenité, D.G.A., De Korte, R. (1986). Electroencephalogr. Clin. Neurophysiol. 63: 35–41. Access no. N45617.
10 de Beukelaar, F., Rutten, R., Tritsmans, L. (1990).Clinical Research Report R 72 063/11. JRF, Access no. N71337.

New Antiepileptic Drugs (Epilepsy Res. Suppl. 3)
Editors: F. Pisani, E. Perucca, G. Avanzini, A. Richens
© 1991 Elsevier Science Publishers B.V. (Biomedical Division)

CHAPTER 16

Losigamone

Ulrich Stein, Klaus Klessing, Shyam S. Chatterjee

Dr. Willmar Schwabe Arzneimittel, Karlsruhe, Germany

Introduction

Synthesis and pharmacological screening of analogs and derivatives of natural five and six-membered lactones occurring in various piper species led to the identification of several tetronic acid derivatives with anticonvulsant activity in animal models. Based on these observations systematic structure activity studies were initiated in the research facilities of Dr. Willmar Schwabe Arzneimittel, Karlsruhe. Losigamone (AO-33, Fig. 1) exhibited strong anticonvulsant activity paired with excellent tolerability and was finally chosen for further development in a joint effort of Dr. Willmar Schwabe Arzneimittel and the Epilepsy Branch of the NINCDS.

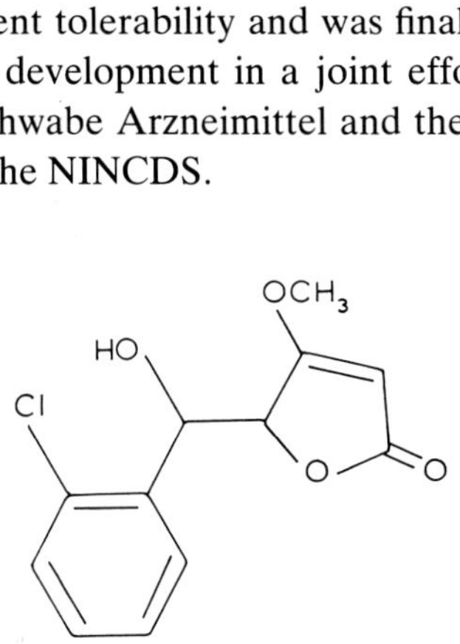

Fig. 1. Chemical structure of AO-33. Formula $C_{12}H_{11}ClO_4$; M. weight 254.67; CA-name (±)-5(R,S)-α(S,R)-5-(2-chlorophenylhydroxymethyl)-4-methoxy-2(5H)-furanone; CA-Reg. No. 112 856-44-7; INN-name Losigamone.

The structure as well as the mechanism of action of losigamone cannot be related to any of the prototype agents or to other anticonvulsants that are currently being developed. This could suggest that losigamone may be useful in controlling seizures that have been uncontrollable so far. Like valproic acid losigamone does not contain nitrogen.

Chemistry

Losigamone, threo-5-(2-chlorphenylhydroxymethyl)-4-methoxy-2(5H)-furanone, is related to the β-methoxybutenolides, which constitute a partial structure of numerous naturally occurring compounds. There is a close relationship between losigamone and the fadyenolides[1] which have been isolated from Piper fadyenii and the piperolides[2] isolated from Piper sanctum, which is especially true for their dihydro-derivatives.

A threo-selective synthesis was developed to facilitate large scale production of the compound in high purity[3]. The structure of losigamone has been verified by means of analysis, IR, ^{1}H-NMR, ^{13}C-NMR and mass-spectroscopy as well as X-ray analysis[4,8].

Losigamone is a racemic mixture of the two possible enantiomers and a neutral white crystalline substance with a melting point of 148° ±3°C which is practically insoluble in water (0.04%). It has no prominent taste. The octanol/pH 7.2 buffer partition coefficient K' at 37°C is 33 and the protein binding in human plasma amounts to 60%. For pharmaceutical usage the medium particle size is ground to less than 15 μm.

Losigamone is assayed in body fluids by means of HPLC and UV-detection.

Pharmacology

The anticonvulsant profile of losigamone was determined by studies conducted by Dr. Willmar Schwabe Arzneimittel[7] and by the Epilepsy Branch of NINCDS[6]. Losigamone is effective when given intraperitoneally and orally in mice, rats and gerbils at non-neurotoxic doses. The degree and duration of protection against convulsive stimuli varies with the species and the type of convulsant. The ED_{50} for losigamone in the maximal electro-shock is 6.7 mg/kg in rats and 19.8 mg/kg in mice. Mice are protected dose dependently against clonic metrazol convulsions whereas rats are not.

In mice tonic convulsions induced by MES, metrazol, bicuculline, nicotine and aminopyridine as well as clonic convulsions induced by metrazol, bicuculline and picrotoxin are suppressed by losigamone in a dose dependent manner. However, there is no protection against tonic strychnine and picrotoxin and clonic NMDA seizures in this species.

The profiles of anticonvulsant activity of losigamone and four prototype agents after oral administration in mice and rats are summarized in Table 1.

Unlike several prototype agents losigamone does not bind to GABA, picrotoxin or benzodiazepine receptors. In high concentrations (9×10^{-4} M) it slightly inhibits (39%) adenosine uptake in mouse brain synaptosomes. In lower concentrations it potentiates GABA-induced chloride in-flux in primary cultured spinal cord neurons. GABA antagonists like picrotoxin or bicuculline block this effect of losigamone[6,9]. These observations indicate that the anticonvulsant activity of losigamone could be mediated via GABA-gated chloride channels.

Brain slice experiments[5] showed that losigamone blocks epileptiform activity induced by adding picrotoxin or omitting magnesium or calcium from the perfusion medium supplying hippocampal slices. Spontaneous epileptiform events, induced by 10 to 30 μM PTX were reduced and finally blocked. Stimulus induced burst discharges were shortened in duration, but not blocked. Extracellular calcium concentration changes and associated slow negative field potentials were diminished in a dose dependent manner. Intracellular recordings revealed no effect of losigamone on resting membrane potential, little effect on input resistance, a small increase in the threshold of action potentials and an abbreviation of stimulus-induced paroxysmal depolarisation shifts (PDS). Spontaneous PDS initially decreased in duration and finally disappeared. In this model losigamone is much more effective than valproic acid. Phenytoin is of the same order of effectiveness.

Toxicology

Acute oral doses of up to 600 mg/kg given to Sprague-Dawley rats and NMRI mice did not result in mortality within the seven days of observation. Sedation, ptosis, ataxia and respiratory depression could be observed 3 to 5 hours after application of high doses. Because there was no mortality the LD_{50} for rats and mice is assumed to be higher than 600 mg/kg. Similarly no mortality was observed in 4 week subchronic toxicity studies in rats and dogs at losigamone doses of 600 mg/kg/day. After 15 days daily administration of losigamone to rats and mice no alterations of the ED_{50}s and hence no reduction of the anticon-

TABLE I

PROFILE OF ANTICONVULSANT ACTIVITY OF ORALLY ADMINISTERED LOSIGAMONE (AO-33) AND SOME PROTOTYPE ANTIEPILEPTIC DRUGS IN MICE AND RATS

Add. no. substance	Time of test (hrs)		TD_{50} (mg/kg)		$MES-ED_{50}$ (mg/kg)		$sc\ Met-ED_{50}$ (mg/kg)	
	Mice	Rats	Mice	Rats	Mice	Rats	Mice	Rats
AO-33	1/4, 1/4, 1/4	1, 1, 4	148.60 (114.06-184.24) [8.09]	384.29 (254.13-500.32) [4.69]	19.8 (18.05-21.36) [18.85] PI 7.47*	6.73 (4.58-9.75) [4.44] PI 57.1*	33.97 (27.04-43.10) [3.99] PI 4.37*	No protection up to 400 PI <0.96*
Phenytoin	2, 2, 2	1/2, 4, 4	86.71 (80.39-96.09) [13.01]	No ataxia up to 3000	9.04 (7.39-10.62) [6.28] PI 9.59*	29.82 (21.92-38.91) [2.82] PI >100*	No protection up to 300 PI <0.29*	No protection up to 400 PI NA*
Phenobarbitone	2, 2, 2	1/2, 5, 5	96.78 (79.88-115.00) [8.51]	61.09 (43.72-95.85) [3.00]	20.09 (14.78-31.58) [5.20] PI 4.82*	9.14 (7.58-11.86) [4.12] PI 6.68*	12.59 (7.99-19.07) [3.84] PI 7.69*	11.55 (7.74-15.00) [4.08] PI 5.29*
Ethosuximide	1/2, 1/2, 1/2	2, 2, 2	879.21 (839.89-933.51) [30.50]	1012.31 (901.66-1109.31) [15.33]	No protection up to 2000 PI <0.44*	No protection up to 1200 PI <0.084*	192.71 (158.59-218.44) [7.39] PI 4.56*	53.97 (45.57-60.85) [9.05] PI 18.8*
Valproate	2, 1, 1	1/2, 1/2	1264.39 (800-2250) [4.80]	280.26 (191.32-352.76) [4.63]	664.80 (605.33-718.00) [18.17] PI 1.90*	489.54 (351.14-728.37) [2.90] PI 0.57*	388.31 (348.87-438.61) [8.12] PI 3.26*	179.62 (146.73-210.35) [8.62] PI 1.56*

( ) 95% confidence interval.
[ ] slope, regression line.
* protective index = TD_{50}/ED_{50}.
NA, not applicable.

vulsant activity of losigamone or other prototype anticonvulsants in different convulsion models were seen. This suggests that there was no development of tolerance or enzyme induction after prolonged treatment.

The administration of doses up to 400 mg/kg losigamone to beagle dogs for 26 weeks did not result in mortality or ophthalmoscopic, electrocardiographic or gross necropsy findings suggestive of compound effect. Clinical observations indicative of compound effect include emesis, salivation, ataxia, languid appearance, prostration and mydriasis. Increases of the mean activated partial thromboplastin time, mean total cholesterol and mean alkaline phosphatase were found. As to organ weight, liver and heart weight increases were noted. Histomorphologic evaluation revealed an increased incidence of uterine hyperplasia and lower ovarian weights. Some of these observations were also seen with 110 mg/kg. Since dosages in this study were 30, 110 and 400 mg/kg, the apparent no-observable-effect level of losigamone in this study is 30 mg/kg.

In a 26 weeks study in rats, doses of 5, 15 and 125 mg/kg were administered by oral gavage. At 125 mg/kg languid behaviour and ataxia, depression of mean body weight gain value in females, increase in total cholesterol in females, relative liver and kidney weight increases, increased incidence and severity of centrolobular hypertrophy, increase in both number and size of ovarian follicles were seen. Some of the effects were also seen at 15 mg/kg. An apparent compound-related change was also noted in the light microscopic examination of the liver of males given 5 mg/kg/day. Therefore, the no-observable-effect level of losigamone in rats is less than 5 mg/kg/day.

Mutagenicity tests using Salmonella typhimurium strains TA 98, 100, 1535, 1537, 1538 and Escherichia coli WP 2 uvr A did not show any evidence that losigamone has mutagenic activity.

Pharmacokinetics

Pharmacokinetics and metabolism of losigamone were studied in mice, rats, dogs and man. There is considerable inter-species variation. However, in none of the species tested losigamone or its metabolites accumulated after repeated dosing. In rats and mice a good correlation of the plasma and brain concentrations to the duration of anticonvulsant action in various models was observed. None of the metabolites identified until now in mouse, rat and man possess anticonvulsant activity.

Losigamone given orally to man reaches its peak concentration at 2 to 3 hours with an absorption half life of approximately 0.6 h. The elimination half life is 4 to 5 h. As shown in Figure 2, losigamone exhibits a good dose/blood level correlation.

When 3 × 500 mg of losigamone were given daily to volunteers for seven days the elimination half life was slightly but not statistically significantly reduced. There were also no changes in the caffeine clearance before and after the seven days of administration.

Data on drug interactions are not yet available for losigamone.

Tolerability in volunteers

In an ascending dose tolerance study losigamone was given in doses of 100 to 1000 mg. It proved to be well tolerated up to the highest dose and no relevant changes in blood pressure, ECG and respiratory parameters were seen. Laboratory parameters did not show changes thought to be related to the trial drug.

Subjective adverse drug reactions were mild. Tiredness and dizziness seemed to increase with dose but they were also seen in the placebo group, as were gastro-intestinal and ophthalmological disturbances.

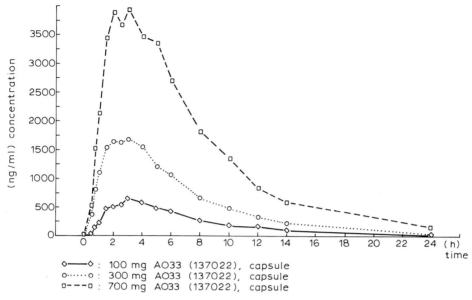

Fig. 2. Median blood concentrations of losigamone after oral administration of three different single oral doses in normal volunteers.

Losigamone was also well tolerated when given to volunteers in doses of 3 × 500 mg for seven days. In a few cases there were slight elevations of liver enzymes. The significance of these changes, however, is unclear because reexposition in one case was negative and another subject showed a positive hepatitis serology.

Phase I will be completed soon and Phase II studies will be started in various European countries and the U.S.A.

References

1   Pelter, A., Al-Bagati, R., Hänsel, R., Dinter, H., Burke, B. (1981) Tetrahedr. Lett. 22: 1545–48.

2   Hänsel, R., Pelter, A. (1971) Phytochem. 10: 1627–34.

3   Chatterjee, S.S., Klessing, K., EP-A2 0247320, DE-A 3615157, US 4,855,320.

4   Pelter, A., Ayoub, M.T. (1981) J.C.S. Perkin Trans I: 1173.

5   Köhr, G. (1989) Dissertation: Analyse von Wirkungsmechanismen potentieller Antikonvulsiva an in-vitro Epilepsiemodellen. Universität Bonn.

6   Swinyard, E.A., Wolf, H.H., Franklin, M.R., White, H.S., Woodhead, J.H., Kupferberg, H.J., Stables, J.P. (1987) The early evaluation of anticonvulsant drugs (Contract No. NO1-NS-4-2361) 'Red Book' NINCDS.

7   Chatterjee, S.S., Klessing, K., Stein, U. (1989) 18th International Epilepsy Congress, New Delhi, India. Book of Abstracts, page 6, No. 23.

8   Klessing, K., Chatterjee, S.S. (1989) Internationales Symposium: Dreidimensionale Struktur und Wirkung, Schliersee, October 2–4.

9   Ticku, M.K. (1989) Publication in preparation.

New Antiepileptic Drugs (Epilepsy Res. Suppl. 3)
Editors: F. Pisani, E. Perucca, G. Avanzini, A. Richens
© 1991 Elsevier Science Publishers B.V. (Biomedical Division)

CHAPTER 17

Oxcarbazepine

Peder Klosterskov Jensen[1], Lennart Gram[2], Marcus Schmutz[1]

[1]*R&D Department, CIBA-GEIGY Ltd., Basle, Switzerland;* [2]*Dianalund Epilepsy Hospital, Dianalund, Denmark*

Introduction

Oxcarbazepine (10,11-dihydro-10-oxo-carba-mazepine) is a new antiepileptic compound belonging to the same chemical class as carbamazepine. The two compounds differ in that only oxcarbazepine has a keto-group in the 10,11 position. This keto group is responsible for the specific biotransformation in humans which differs significantly from that of carbamazepine and other antiepileptic compounds.

Oxcarbazepine has been developed as a follow-up compound for carbamazepine with a similar efficacy profile, but with a tolerability profile superior to that of carbamazepine. In man, oxcarbazepine is almost immediately and completely metabolised to the 10-monohydroxy derivate, 10,11-dihydro-10-hydroxy carbamazepine (MHD). The antiepileptic effects of oxcarbazepine and MHD are comparable in both animal and man.

Chemistry

Oxcarbazepine is a derivative of the dibenz[b,f]azepine series. With carbamazepine it

Compound	Oxcarbazepine	GP 47 779	Carbamazepine
Structure	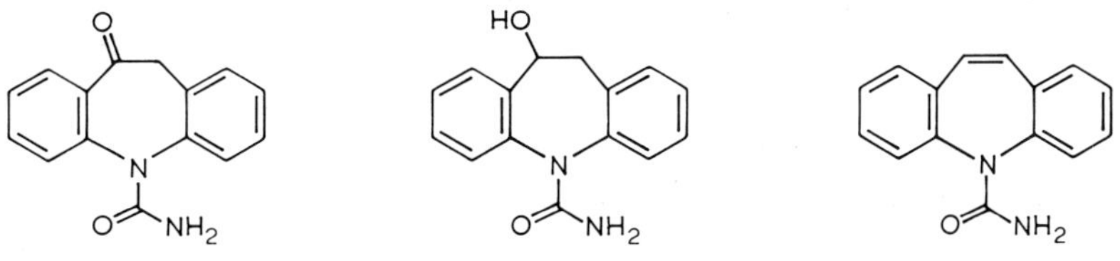		
Molecular weight	252.27	254.29	236.27
Solubility [1]	0.13 g [g/1]	4.2 [g/1]	0.24 [g/1]
Partition coefficient [2]	20.4	8.8	57.9

Fig. 1. Structural formulae and physicochemical properties.
[1] Solubility in water, 37° C.
[2] For partition between n-octanol/aqueous buffer pH 7.4, 25° C.

shares the tricyclic skeleton but is structurally different in the 10,11 position (Fig. 1). The molecular weight of oxcarbazepine is 252.27. From its physicochemical properties it can be classified as a neutral lipophilic compound with a very low solubility in aqueous media.

Animal pharmacology

The information in this section is based partly on unpublished data from tests conducted by or for CIBA-GEIGY Ltd., Basle, Switzerland, where all reports are on file.

Electroshock and pentylenetetrazole tests in mice and rats

Oxcarbazepine and MHD inhibit the hind-limb extension after a supramaximal electroshock in both rats and mice. The ED_{50} is 10–20 mg/kg p.o.[1]. This effect lasts for about 8 hours in the rat. No loss of efficacy was noted in rats treated with oxcarbazepine for four weeks[2]. In the pentylenetetrazole test in mice, both oxcarbazepine and MHD were found to be less active than in the electroshock test (oxcarbazepine ED_{50} 23–30 mg/kg p.o., MHD ED_{50} 52 mg/kg p.o.)[1].

Picrotoxin and strychnine tests in mice

Both these tests are based on antagonism of the inhibitory neurotransmitters, GABA and glycine. In both tests, the anticonvulsive activity of oxcarbazepine and MHD is relatively weak, with ED_{50} values in the range of 150 to 250 mg/kg p.o.[1]

Chronic aluminum foci in rhesus monkey

This model is considered to be a model for partial and post-traumatic seizures in man[3,4]. In this model oxcarbazepine, after a single-dose, showed complete seizure suppression at 50 mg/kg p.o. and very marked reduction of seizure activity at 20 mg/kg i.m. MHD was less effective, as seizures were not completely abolished following 50–100 mg/kg p.o. and less markedly inhibited at 20 mg/kg i.m.

In multiple-dose studies, daily doses of oxcarbazepine, 30 to 60 mg/kg i.m. for 7 days, abolished or markedly reduced seizures. After p.o. administration, it was less potent, as a dose of 100 mg/kg was needed to reduce seizure frequency by 50 to 99%. Also in this test, MHD was found to be slightly less effective.

The animal tolerability of oxcarbazepine and MHD has been evaluated with regard to CNS side effects, such as decreased motility, ataxia and muscular hypotonia, in mice and rats. No effects were seen in doses below 100 mg/kg p.o. in rats. The threshold doses for sedation in mice are considered to be 30 mg/kg p.o. and 100 mg/kg p.o. for oxcarbazepine and MHD, respectively[1]. In dogs and tupaias oxcarbazepine does not lead to unwanted effects in doses up to 300 and 200 mg/kg p.o., respectively.

From the above data it seems to be justified to assume that oxcarbazepine and MHD will be effective and well tolerated in patients suffering from generalised tonic-clonic seizures and partial seizures with and without secondary generalisation.

Clinical pharmacokinetics

Following peroral administration to humans, oxcarbazepine is virtually completely absorbed, as 96% of a single dose of 400 mg ^{14}C-oxcarbazepine was excreted with the urine within 10 days (85% within 48 hours)[5]. However, oxcarbazepine reaches only very low concentrations in plasma because of its rapid biotransformation to the antiepileptically active metabolite, MHD[6,7]. MHD is therefore the molecule responsible for the antiepileptic efficacy in humans. MHD is formed by a metabolic reduction of the keto group of oxcarbazepine and is thereafter inactivated by conjugation with glucuronic acid (Figure 2). A minor

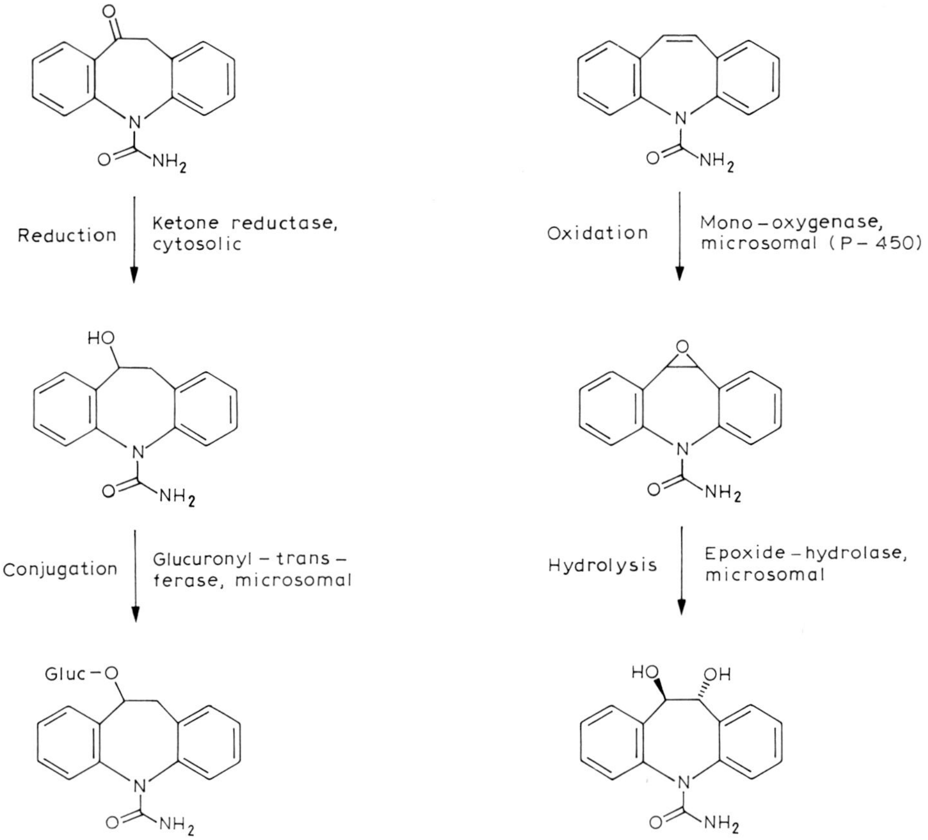

Fig.2. Major metabolic pathways of oxcarbazepine and carbamazepine in man.

metabolic pathway is an oxidation to 10,11-dihydro-10,11-trans-dihydroxy carbamazepine[8]. The latter metabolite only constitutes a very small percentage of the AUC of MHD. In comparison to carbamazepine (primarily cleared by oxidative processes), this means that oxcarbazepine cannot give rise to the formation of carbamazepine-10,11-epoxide[7] which has been suspected to contribute to the neurotoxicity of carbamazepine[9,10] (Fig. 2). The kinetics of oxcarbazepine and MHD in humans are therefore largely controlled by enzymes which do not involve the cytochrome P-450 enzymes.

Because of the rapid and almost complete bio-

transformation of oxcarbazepine to MHD, all pharmacokinetics in humans have been based on this metabolite.

The elimination half-life of MHD in healthy volunteers is 8–10 hours after both single-dose and multiple-dose administration (no autoinduction). The elimination was found to be mono-exponential. Dose linearity up to 2700 mg has been shown in patients under steady-state conditions.

Based on the different biotransformation compared with carbamazepine, one can assume that the interaction potential of oxcarbazepine is much less than that of other antiepileptic compounds metabolised via cytochrome P-450 depen-

138

dent enzymes (see Drug Interactions).

Activity and tolerability in epileptic patients

A total of 791 patients have been included in the clinical trial programme. Of these, 581 patients were exposed to either oxcarbazepine or its primary metabolite, MHD. 326 patients were treated in monotherapy. The duration of treatment varied from 8 to 52 weeks, 143 patients were treated for more than one year. Only the pivotal studies in monotherapy will be addressed in this presentation.

Therapeutic efficacy

The therapeutic efficacy of oxcarbazepine has been evaluated in two double-blind controlled studies in monotherapy[11,12], in a total of 353 patients suffering from primary generalised seizures (only tonic-clonic) and/or partial seizures with or without secondary generalisation, classified according to the International Classification of Epilepsy Seizures[13].

The first study[11] included 235 patients, all with newly diagnosed, previously untreated epilepsy. All patients were randomly allocated to treatment with oxcarbazepine (starting dose 300 mg) or carbamazepine (starting dose 200 mg).

The daily dose was then individually titrated according to efficacy and tolerability. The duration of the titration phase was between 4 and 8 weeks. After determination of the individual optimal dose this dose was maintained for a further 12 weeks. In patients who responded well to the treatment and were willing to continue the study the treatment was continued for a further 36 weeks. No statistically significant difference in the number of seizures per month was observed between the two treatment-groups (oxcarbazepine, mean: 0.4 ± 3.0; carbamazepine, mean: 0.3 ± 1.4) during the maintenance phase. The mean daily doses of the trial medication were 1040 mg

(range 300–1800 mg) in patients treated with oxcarbazepine and 684 mg (range 300–1400 mg) in patients treated with carbamazepine. The design of the second study[12] was similar to the first with the only exception being that the inclusion criteria for this study allowed patients not well controlled under previous antiepileptic treatment. Also in this study, no difference in the mean number of seizures per month was observed (oxcarbazepine, mean: 1.0 ± 2.1; carbamazepine, mean: 1.0 ± 2.5).

The therapeutic effect of oxcarbazepine, in monotherapy, compared with that of its primary metabolite MHD, was evaluated in 59 patients. No difference in the mean number of seizures was observed between the two treatment-groups (unpublished data).

From the clinical studies performed until now it can be concluded that oxcarbazepine is as effective as carbamazepine in the treatment of generalised tonic-clonic seizures and partial seizures with and without secondary generalisation[11,12]. The equivalent effective dose is about 50% higher than that of carbamazepine. Although all pivotal studies have been performed in a t.i.d. dosage regimen, preliminary results indicate that a b.i.d. dosage regimen may be possible in some patients.

Tolerability

The tolerability of oxcarbazepine has been studied on the basis of the effect on a number of laboratory parameters (white blood cells, red blood cells and platelet counts, liver function tests and urine analysis) as well as pulse rate, blood pressure and side effects (reported by the patients, as well as by the treating physicians). Although a trend toward a better overall tolerability profile of oxcarbazepine, compared with carbamazepine, was observed (percentage of patients with side effects: oxcarbazepine 68%, carbamazepine 74%; mean number of side effects per patient: oxcarbazepine 2.8, carbamazepine 3.5), this difference was not statistically significant[11]. However, looking at the number of side effects necessitating withdrawal of the treatment in the largest controlled study in monotherapy, a statistically significant difference was observed in favour of oxcarbazepine[11]. No difference in the nature of side effects was observed between the two treatment-groups. The question of allergic skin rashes during treatment with oxcarbazepine has been studied by Jensen et al[17]. In this study 51 patients with allergic skin rashes during treatment with carbamazepine were changed to oxcarbazepine. Only 27% of the patients experienced cross-allergy between carbamazepine and oxcarbazepine. Recent information on a lowering of serum sodium[14,15,16] during treatment with oxcarbazepine still needs to be investigated in larger controlled studies.

Drug interactions

The metabolic characteristics of oxcarbazepine and MHD (not depending on cytochrome P-450 enzymes) makes the risk of metabolic drug interactions very low, compared with a large proportion of the established antiepileptic compounds.

In order to prove this low risk of drug interactions, a large interaction programme has been started in patients and volunteers. Three studies from this programme have already been completed. The results show that the pharmacokinetics of oxcarbazepine and MHD were not influenced by either propoxyphene, cimetidine or erythromycin (unpublished data), all compounds that have been shown to inhibit the biotransformation of carbamazepine[18,19,20].

Acknowledgements

We thank Dr. J.W. Faigle, R&D Department, CIBA-GEIGY Ltd., Basle, Switzerland, for his contribution to the section on 'Clinical Pharmacokinetics'.

References

1 Baltzer, V., Schmutz, M. (1978) In: Meinardi, H., Rowan, A.J. (eds.), Advances in Epileptology. Swets und Zeitlinger, Amsterdam/Lisse, pp. 295–299.
2 Schmutz, M., David, J., Grewal, R.S., Bernasconi, R., Baltzer, V. (1986) In: Frey, H. et al. (eds.), Tolerance to Beneficial and Adverse Effects of Antiepileptic Drugs. Raven Press, New York, pp. 25–34.
3 David, J., Grewal, R.S. (1977) Life Sci. 21: 1109–1116.
4 David, J., Grewal, R.S. (1976) Epilepsia 17: 415–422.
5 Feldmann, K.F., Brechbühler, S., Faigle, J.W., Imhof, P. (1978) In: Meinardi, H., Rowan, A.J. (eds), Advances in Epileptology. Swets and Zeitlinger, Amsterdam/Lisse, pp. 290–294.
6 Anonymous (1986) Drugs of the Future 11: 844–847.
7 Faigle, J.W., Menge, G.P. (1990) International Clinical Psychopharmacology 5 (Supplement 1): 73–82.
8 Schütz, H., Feldmann, K.F., Faigle, J.W., Kriemler, H.P., Winkler, T. (1986) Xenobiotica 16: 769–778.
9 Patsalos, P.N., Stephenson, T.J., Krishna, S., Elyas, A.A., Lascelles, P.T., Wiles, C.M. (1985) Lancet 2: 496.
10 Gilham, R.A., Willias, N., Weidmann, K., Butler, E., Larkin, J.G., Brodie, M.J. (1988) J. Neurol. Neurosurg. Psychiatry 51: 929–933.
11 The Scandinavian Oxcarbazepine Study Group (1989) Epilepsy Res. 3: 70–76.
12 The Scandinavian Oxcarbazepine Study Group (1990) Acta Neurol. Scand. (submitted).
13 Commission on Classification and Terminology of the International League Against Epilepsy (1981) Epilepsia 22: 489–501.
14 Houtkooper, M.A., Lammertsma, A., Meyer, J.W.A., Goedhart, D.M., Meinardi, H., van Oorschot, C.A.E.M., Blom, G.F., Höppener, R.J.E.A., Hulsman, J.A.R.J. (1987) Epilepsia 28: 693–698.
15 Johannessen, A.C., Nielsen, O.A. (1987) Epilepsy Res. 1: 155–156.
16 Nielsen, O.A., Johannessen, A.C., Bardum, B. (1988) Epilepsy Res. 2: 269–271.
17 Jensen, N.O. (1985) 16th International Epilepsy Congress, Hamburg.
18 Dam, M., Christensen, J.M., Brandt, J., Hansen, B.S., Hvidberg, E.F., Angelo, H., Lous, P. (1980) In: Johannessen SI et al. (eds.), Antiepileptic Therapy: Advances

140

in Drug Monitoring, Raven Press, New York, pp. 299–306.

19 Dalton, M.M., Powell, J.R., Messenheimer, J.A. (1985) Drug Intell. Clin. Pharm. 19: 456.

20 Barzaghi, N., Gatti, G., Crema, F., Monteleone, M., Amione, C., Leone, L., Perucca, E. (1987) Br. J. Clin. Pharmacol. 24: 836–838.

New Antiepileptic Drugs (Epilepsy Res. Suppl. 3)
Editors: F. Pisani, E. Perucca, G. Avanzini, A. Richens
© 1991 Elsevier Science Publishers B.V. (Biomedical Division)

CHAPTER 18

Ralitoline

Henning Anhut, Gerhard Satzinger, Albrecht von Hodenberg

Gödecke AG, Research Institute, Freiburg, Germany

Introduction

Ralitoline, a structurally novel anticonvulsant, was a chemical development originating from the 2-methylene-thiazolidinone-4 moiety known to modify membrane ion transport mechanisms[1]. Based on its pharmacological properties, ralitoline is expected to be a potent and broad-spectrum antiepileptic drug[2,3].

Chemistry

Ralitoline, (CI-946; (Z)-N-(2-chloro-6-methylphenyl)-2-(3-methyl-4-oxo-2-thiazolidinylidene) acetamide (Figure 1) is sparingly soluble in water (0.04%) and has a partition coefficient (Log P) in n-octanol/buffer of 1.72 at pH 7.4.

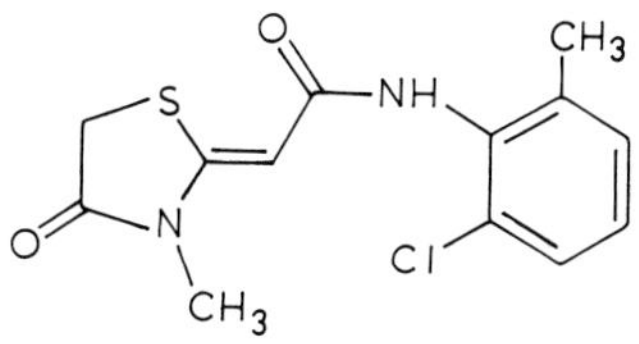

Pharmacological properties

Electrically and chemically induced tonic convulsions

In the maximal electroshock seizure (MES) model in rats and mice, maximal anticonvulsant activity occurred 15 to 30 minutes after oral administration (p.o.); ED_{50} values determined with a 30-minute premedication time were 7 and 3 mg/kg p.o. in male mice and rats, respectively. The onset of anticonvulsant protection was within 6 minutes and lasted for up to 8 hours in rats de-

Fig. 1. Chemical structure of ralitoline.

pending on the dose of ralitoline. Ralitoline was approximately three to six times more potent than phenytoin and carbamazepine in the MES model. In the MES model in mice pretreated intravenously (i.v.) with ralitoline at different times (1, 15, and 30 minutes), ED_{50} values were 1.5, 5.8, and 6.8 mg/kg respectively.

Ralitoline also protected mice from tonic extension induced by several chemoconvulsants, such as 3-mercaptopropionic acid, semicarbazide, isoniazid, bicuculline, picrotoxin, strychnine, and pentylenetetrazol (PTZ). Ralitoline ED_{50} values versus the various chemoconvulsants were in the range of 0.7 to 5.0 mg/kg p.o. Similar or higher values were found for carbamazepine (2.4 to 13.7 mg/kg p.o.) phenytoin (3.9 to 37.7 mg/kg p.o.) and valproate (71 to 251 mg/kg p.o.).

Pentylenetetrazol induced clonic seizures

The threshold test for clonic convulsions after i.v. infusion of PTZ in mice is used to predict drugs effective against generalized seizures of the absence (petit mal) type. An elevation of the convulsant threshold to 120% of control values was obtained in mice treated with 4.5 mg/kg p.o. of ralitoline 30 minutes prior to PTZ infusion (10 mg/ml, 0.15 ml/min); maximal effects were observed with 6 mg/kg (152%) and 6.5 mg/kg p.o. (176%). Valproate increased the convulsant threshold by 120% at a dose of approximately 100 mg/kg p.o., while phenytoin and carbamazepine were inactive in this model.

In another absence seizure model, the less sensitive subcutaneous (s.c.) PTZ threshold test model, ralitoline was inactive up to 40 mg/kg p.o.

Hippocampal kindling model

Male rats were electrically kindled in the ventral hippocampus and the threshold for afterdischarge was determined. The electrical current necessary to produce a focal afterdischarge increased with increasing doses of ralitoline. A dose of 30 mg/kg p.o. induced a six- to eightfold increase over the baseline threshold, suggesting that ralitoline may be effective for the treatment of partial seizures.

Genetic animal models with reflex seizures

In the reflex seizure model with selectively inbred male Mongolian gerbils, ralitoline only protected against tonic-clonic seizures with ED_{50} values between 10 to 30 mg/kg depending on premedication time (15, 30, or 60 min). Myoclonic seizures were not influenced by ralitoline or by phenytoin while carbamazepine, phenobarbitone and valproate were active against all seizure types[4].

Ralitoline was also tested in male audiogenic seizure-susceptible DBA/2 mice. Sound-induced wild running phase/clonic convulsions and tonic extensions were prevented by 30-minute pretreatment with ralitoline with ED_{50} values of 2.9 and 2.3 mg/kg p.o., respectively.

Intermittent photic stimulation of photosensi-

tive baboons is suggested as a model for photomyoclonic seizures and myoclonic absence epilepsy[5]. Ralitoline reduced the severity of the myoclonic response to photic stimulation in a dose dependent manner, 5 minutes after i.v. doses of 0.3, 1.0, and 3.3 mg/kg. Following administration of 5 mg/kg p.o., ralitoline showed a pronounced sustained anticonvulsant effect over 4 hours. This dose did not produce any sign of neurological toxicity, such as sedation, ataxia, muscle relaxation, or nystagmus. These data suggest that ralitoline may be effective for the treatment of myoclonic epilepsy.

Anticonvulsant efficacy and neurotoxicity evaluated by NINDS

Ralitoline was also investigated in the NINDS Anticonvulsant Screening Project, which includes the MES test, the s.c. PTZ seizure threshold test, and the Rotorod minimal neurotoxicity test. Ralitoline was ineffective in the s.c. PTZ test in mice and rats at doses up to 300 and 600 mg/kg p.o., respectively (see above). The results of the MES and the neurotoxicity test are summarized in Table 1.

The protective index (PI = TD_{50}/ED_{50}) values calculated for ralitoline are similar to or substantially higher than those of phenytoin, valproate and carbamazepine (mice: 9.59, 1.9, 14.1[*]; rats: >100, 0.57, 95.7[*] respectively; [*]from Ref. 6; see Table 1).

Minimum effective dose and minimum effective plasma concentration of ralitoline in the electroconvulsive threshold model[7]

In mice, the dose that increases the electroconvulsive threshold by 2 mA at the time of maximum anticonvulsant activity of the test drug was determined and the corresponding minimum effective plasma concentration was measured at the same time.

Maximal prevention of tonic extension oc-

TABLE 1

ANTICONVULSANT ACTIVITY AND NEUROTOXICITY OF RALITOLINE. NINDS SCREENING PROJECT

Pretreatment time (min)	MES test ED_{50} (mg/kg p.o.)	Rotorod test TD_{50} (mg/kg p.o.)	PI TD_{50}/ED_{50}	
Mice	15	15.4 (11.8–21.8)	205 (143–311)	13.3
Rats	60	1.42 (1.15–1.62)	574 (349–823)	404

Mean TD_{50} and ED_{50} values (mg/kg) with corresponding 95% confidence intervals in parentheses

curred 2 minutes after intraperitoneal (i.p.) administration of ralitoline. The minimum effective dose of ralitoline required to increase the electroconvulsive seizure threshold by 2 mA was 0.97 mg/kg i.p.; the corresponding mean minimum effective plasma concentration of ralitoline was 0.34 μg/ml (range 0.08 to 0.54 μg/ml, n=21). The minimum effective dose and plasma concentration of ralitoline in mice were lower than those of other standard antiepileptic drugs except for benzodiazepines[8].

Subchronic treatment with ralitoline

A subchronic study with ralitoline was performed using the MES model. The drug was administered to male mice for 2 or 14 days twice daily at a dose of 6 mg/kg p.o. The ED_{50} value in animals treated for 2 days was 6.5 mg/kg as compared to 5.7 mg/kg in controls receiving ralitoline only once. The ED_{50} value in mice treated for 14 days was 8.9 mg/kg, showing that no marked tolerance developed to the anticonvulsant effect.

Biochemical pharmacology

Ralitoline did not interfere with the GABA/benzodiazepine receptor complex. At concentrations between 10 and 50 μM, ralitoline had no influence on GABA-uptake. Furthermore, ralitoline (5 mg/kg p.o.) had no effect on the GABA content in mouse nerve terminals ex vivo.

Ralitoline displaced binding of ^{3}H-batrachotoxinin from rat brain synaptosomes with an ap-

parent K_d of 25 μM, indicating interaction with voltage-dependent sodium channels. Electrophysiologic studies indicated that ralitoline blocked sustained repetitive firing in cultured mouse spinal cord cells at concentrations between 0.3 and 3.4 μM; no effect on postsynaptic responses were observed. These results also suggest that ralitoline causes a cumulative inactivation of sodium channels[9]. Similar effects were observed in the guinea pig papillary muscle preparation. Both ralitoline and phenytoin reduced the action potential upstroke velocity, suggesting an inhibitory effect on voltage-dependent sodium channels[10]. Thus, the anticonvulsant effect of ralitoline may be mediated by an impairment of sodium fluxes into neuronal cells as postulated for phenytoin[11].

Pharmacokinetics

Ralitoline is well absorbed. In man, 90% of a radioactive dose was recovered in urine[12]. After oral administration, maximum ralitoline plasma concentrations were achieved approximately 1 hour postdose in rats and dogs and 2.8 hours postdose in man, and declined with elimination half-lives of approximately 2 hours in rats and dogs and 4.5 hours in man.

Ralitoline crosses easily the blood brain barrier. The ratios of brain to blood concentrations remain constant for up to 2 hours postdose. Binding of ralitoline to human plasma proteins and human serum albumin was 74% and 45% respectively.

Ralitoline is extensively metabolized in rats, dogs, and man, with less than 1% of the dose excreted in urine as unchanged drug. Major routes of biotransformation are hydroxylation of the aromatic ring, oxidation of the methyl group at the aromatic ring, and sulfur oxidation at the thiazolidinone ring. All three routes are followed by conjugation reactions with glucuronic acid and/or sulphuric acid. Only the unchanged drug is responsible for the pharmacological activity[12].

Dose proportionality studies were carried out following oral administration of ralitoline at single doses of 10 to 150 mg[12] and at t.i.d. doses of 50 to 300 mg in healthy male subjects.

Following single dose administration, peak plasma concentrations (Cmax) and $AVC_{0-\infty}$ values increased in proportion to dose[12]. Following multiple dose administration, no significant differences in dose normalized pharmacokinetic parameters were found (50 to 200 mg, Table 2). Elimination kinetics was unaltered following repeated drug administration over 12 days.

Activity studies in epileptic patients

An early double-blind placebo-controlled phase II study investigated the effect of ralitoline on focal spike-wave activity in 10 refractory patients with localization-related epilepsy[14]. Five out of seven evaluable patients had a plasma level-related transient fall in spike-wave activity after oral administration of ralitoline (70 mg) in addition to their current antiepileptic medication. One patient on placebo also responded and one other patient exhibited no difference in response to ralitoline or placebo. Three were unevaluable due to insufficient spike wave activity for comparison during study observation time. It was concluded from this study that ralitoline is active against human epileptic EEG spike activity.

Drug interactions

To determine whether phenytoin or phenobarbitone alter ralitoline pharmacokinetics and/or whether ralitoline administration alters the steady-state plasma concentrations of these drugs, ralitoline was coadministered to healthy male subjects with phenytoin or phenobarbitone, respectively[13].

Ralitoline single dose Cmax and $AUC_{0-\infty}$ values were approximately two-fold higher when ralitoline was given alone than when it was coadministered with phenytoin or phenobarbitone.

TABLE 2

RALITOLINE PHARMACOKINETIC PARAMETERS FOLLOWING SINGLE AND MULTIPLE (T.I.D.) ORAL DOSES OF 50, 100, AND 200 MG IN HEALTHY MALE VOLUNTEERS

Pharmacokinetic parameter		50 mg	100 mg	200 mg
Peak concentration [ng/ml]	SD	545 ± 209	646 ± 331	1170 ± 233
	MD	628 ± 207	931 ± 387	1628 ± 691
Time of peak [h]	SD	2 ± 1	2 ± 1	3 ± 1
	MD	2 ± 1	2 ± 1	2 ± 1
Half-life* [h]	SD	4.6	4.7	5.9
	MD	4.2	5.5	4.1
$AUC_{0-\infty}$ [µg·h/ml]	SD	3.66 ± 1.67	6.29 ± 4.02	10.3 ± 4.00
AUC_{0-8} [µg·h/ml]	MD	3.50 ± 1.56	5.62 ± 2.66	9.50 ± 5.33

Mean values ± standard deviation; SD: single dose; MD: multiple dose; * harmonic mean.

Phenytoin or phenobarbitone pretreatment enhanced the elimination of ralitoline, most likely due to enzyme induction. Changes in terminal elimination half-lives were not of clinical relevance. Linear pharmacokinetics was demonstrated for ralitoline following single and repeated doses during phenytoin or phenobarbitone coadministration. Eleven or fourteen days of ralitoline coadministration with phenytoin or phenobarbitone appeared not to alter steady-state pharmacokinetics of phenytoin or predose phenobarbitone plasma concentrations[13].

References

1 Satzinger, G. (1978) Liebigs Ann. Chem. 473–511.

2 Bartoszyk, G.D., Dooley, D.J., Ganser, V., Satzinger, G. (1986) Naunyn Schmiedeberg's Arch. 332 Suppl. R88, 350.

3 Bartoszyk, G.D., Dooley, D.J., Fritschi, E., Satzinger, G. (1986) In: Meldrum, B.S., Porter, R.J. (eds.), Current Problems in Epilepsy IV: New Anticonvulsant Drugs. J. Libbey, London, pp. 309–311.

4 Bartoszyk, G.D., Hamer, M. (1987) Pharmacol. Res. Commun. 19: 429–440.

5 Naquet, R., Meldrum, B.S. (1972) In: Purpura, D.P., Penry, J.K., Tower, D., Woodbury, Walter, R. (eds.), Experimental Models – a Manual for the Laboratory Worker. Raven Press, New York, pp. 373–406.

6 Gladding, G.D., Kupferberg, H.J., Swinyard, E.A. (1985) In: Frey, H.H., Janz, D. (eds.), Antiepileptic Drugs. Berlin, Springer Verlag, pp. 341–347.

7 Löscher, W., Schmidt, D. (1988) Epilepsy Res. 2: 145–181.

8 Von Hodenberg, A., Löscher, W., Nolting, B., Faßbender, C.P., Fecht-Kempter, I., Taylor Vollmer, K.-O. (1990) Naunyn Schmiedeberg's Arch. Pharmacol. 341, Suppl. R 99, No. 396.

9 Taylor, C.P., Rock, D.M., McLean, M.J., Macdonald, R.L. (1987) Soc. Neurosci. 94, 29.11.

10 Wagner, B., Strumpf, G., Bartoszyk, G. (1986) Pharmacol. Res. Commun. 19: 591–596.

11 Wilow, M. (1986) Trends in Neuroscience 9: 147–149.

12 Von Hodenberg, A., Katzinski, L., Fecht-Kempter, I., Vollmer, K.-O. (1987) In: Aiache, J.M., Hirtz, J. (eds.), Proc. Third Europ. Congress of Biopharmaceutics and Pharmacokinetics, Vol. III, pp. 460–470.

13 Von Hodenberg, A., Fecht-Kempter, I., Vollmer, K.-O., Schleyer, I., Bockbrader, H.N., Sedman, A. (1989) Eur. J. Clin. Pharmacol. 36, Suppl. A270, PP 11.51.

14 Wolf, P., Hubberts, L., von Hodenberg, A., Katzinski, L., Schmidt, B. (1989) In: Manelis, J., Bental, E., Loeber, J.N., Dreifuss, F.E. (eds.), Advances in Epileptology. Raven Press, New York, 17: 212–214.

New Antiepileptic Drugs (Epilepsy Res. Suppl. 3)
Editors: F. Pisani, E. Perucca, G. Avanzini, A. Richens
© 1991 Elsevier Science Publishers B.V. (Biomedical Division)

CHAPTER 19

Remacemide

Keith T. Muir and Gene C. Palmer

Fisons Pharmaceuticals, Divisional Research and Development, Rochester, NY, USA

Introduction

Remacemide hydrochloride (FPL 12924AA) is a diphenyl-ethyl-acetamide derivative which has demonstrated anticonvulsant activity in animal models of epilepsy. The drug protects against maximal electroshock seizures but not pentylenetetrazol-induced seizures in rodents, suggesting potential clinical usefulness against partial and generalized tonic/clonic seizures. The mechanism of action of remacemide is unknown. In vitro studies suggest that the drug does not act via the same mechanisms as currently marketed antiepileptic drugs. The most recent studies indicate that inhibition of the NMDA receptor complex may play a central role in the drug's activity.

Acute safety studies with remacemide in animals have shown the drug to be comparable to others in the class, while early tolerance studies in normal volunteers and epileptic patients showed the drug to have a fairly benign side effect profile at maximally tolerated doses. No serious drug related adverse events have been observed to date. Remacemide is currently in Phase II clinical trials.

Chemistry

Remacemide hydrochloride {(±) 2-Amino-N-(1-methyl-1,2-diphenylethyl)-acetamide mono-hydrochloride} is a diphenyl-ethyl-acetamide derivative. It is the anhydrous monohydrochloride salt of a base with a pK of 8.0. The molecular weight of the drug is 304.82 and its empirical formula is $C_{17}H_{21}ClN_2O$. It is soluble in most aqueous media, e.g. water for injection –40 g/L; normal saline –22 g/L; dilute hydrochloride acid –26 g/L and in ethanol 24 g/L. The compound contains one asymmetric carbon atom and has been resolved into its constituent enantiomers FPL 14144AJ(+) and FPL 14145AJ(−) for further preclinical study.

Chromatographic methods have been developed for the analysis of parent compound and des-

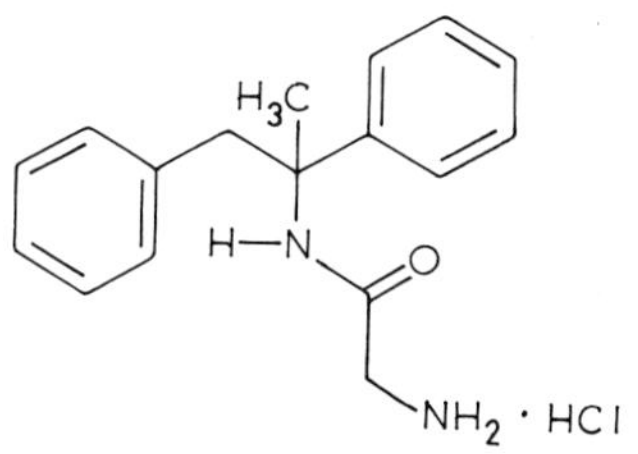

Fig. 1. FPL 12924AA*

---

* Hydrochloride salts are designated AA, fumarate salts AJ, maleate salts AE and the respective free bases XX. Throughout this document, actual quantities of FPL 12924AA dosed (ED50s, LD50s, IC50's etc), will be expressed in terms of the amounts of base (FPL 12924XX) administered: 1.136 mg of the hydrochloride salt corresponds to 1.00 mg of base. Actual quantities of enantiomers dosed will also be expressed in terms of the amounts of base administered.

glycinated metabolite in plasma in the presence of concomitant antiepileptic medications. These methods have been extended to the analysis of parent compound, desglycinated metabolite, hydroxylated metabolites and their conjugates in plasma and urine.

Preclinical pharmacology

Pharmacological testing in mice followed the procedures of Porter et al.[1]. These procedures resulted in the identification of remacemide as the most active in a series of diphenylethylacetamide derivatives in the selective suppression of maximal electroconvulsive shock (MES) seizures. Table 1 lists the comparative activities of remacemide in two rodent species. Analogous data are given for reference agents that are clinically used to treat generalized tonic/clonic seizures. The (+) enantiomer of remacemide (FPL 14144AJ) is less potent than the racemate and the (−) enantiomer (FPL 14145AJ) is more potent than the racemate when evaluated in the MES test. The ED50 dose maintained protection against MES at 50% for 4 hours. This was longer than phenytoin, phenobarbitone, carbamazepine or valproate at their respective ED50 doses. Eighty percent protection against MES was maintained for 8 hours after 3 × ED50 doses of remacemide. This was similar to phenobarbitone but longer than phenytoin, carbamazepine or valproate.

The therapeutic index (TI) was assessed as the ratio of the TD50 for neural impairment (inverted screen test in mice or gang plank escape test in rats) to the ED50 for MES protection. Remacemide compared favourably with the other agents tested (Table 1).

A significant metabolite of remacemide is the desglycinated derivative (FPL 12495AA). FPL 12495AA is more potent than the parent compound against MES but also produced neural impairment at lower doses. The therapeutic index

(TD50/ED50) of FPL 12495AA is lower than that of the parent compound (Table 1).

Further testing for efficacy with remacemide revealed essentially no anti-seizure activities against convulsions elicited by picrotoxin, pentylenetetrazol, strychnine or bicuculline. Neither remacemide nor its desglycine metabolite possessed significant affinity for adenosine-1, GABAa, muscarinic, adenosine uptake, benzodiazepine or strychnine-insensitive glycine receptor sites in vitro. However, remacemide was found to inhibit convulsions induced by N-methyl-D-aspartic acid (NMDA), a property not readily shared with the reference compounds except MK801 (see Table 1) and FPL 12495AA was almost twice as potent as the parent compound.

Receptor binding studies in rat synaptosomes[2] demonstrated that the IC50 for displacement of ^{3}H-MK801 was 68 μM for remacemide, 0.5 μM for FPL 12495AA and 0.014 μM for MK801. Other MK801-like properties were not apparent with remacemide or FPL 12495AA. Unlike MK801, phenobarbitone or valproate, neither remacemide nor FPL 12495AA prevented the development of kindled seizures or acted to inhibit established kindled seizures elicited in rats via subthreshold corneal stimulation[3]. In behavioural tests remacemide, its enantiomers and FPL 12495AA exhibited a tendency to increase spontaneous motor activity at large doses (Table 1). This event might be related to proconvulsant properties of these compounds in which a shortening of the time to clonus and first twitch was evident following iv administration of pentylenetetrazol[4].

In vitro electrophysiological studies with remacemide have demonstrated no modification of evoked synaptic response or penicillin-induced bursting in isolated rat hippocampal slices. However, both remacemide and FPL 12495AA limited sustained repetitive firing in cultured mouse neuroblasts[6]. The ED50 values for remacemide

TABLE I

EFFICACY/SAFETY PROFILES FOR REMACEMIDE, FPL 14144AJ(+), FPL 14145AJ(−), FPL 12495AA(±), AND REFERENCE COMPOUNDS IN RODENTS

Species & Compound	ED50 MES[1]	TD50[2]	TI[3]	ip ED50 NMDA[4]	ip ED50 MES
		Dose (mg/kg)			
CF1 Mice (Harlan)					
Remacemide	48.0	370	7.7	57.4	21.5
FPL 14144AJ(+)	94.2	650	6.9	>75.0	38.7
FPL 14145AJ(−)	29.9	620	20.7	77.2	27.8
FPL12495AA(±)	40.0	98	2.4	32.5	17.1
MK801(±)	1.4	0.9	0.6	1.1	0.4
CF1 Mice (Charles River)					
Remacemide	33.0	581	17.6	−	23.0
Phenytoin	11.0	608	57.4	IA	9.5
Carbamazepine	13.0	137	10.2	IA	11.7
Valproate	631.0	>2000	>3.0	IA	172.0
Phenobarbitone	20.0	105	5.1	IA	22.0
Rats (Sprague Dawley)	ED50 MES[1]	TD50[2]	TI	Kindling[5]	SMA[6]
Remacemide	21.4	847	39.6	IA	↑ 200
FPL 14144AJ(+)	33.0	~1000	~30.3	−	>400
FPL 14145AJ(−)	20.0	529	26.5	−	↑ 100
FPL12495AA(±)	10.0	152	15.2	IA	↑ 100
Phenytoin	13.3	>2000	>150.0	IA	↓ 400
Carbamazepine	4.0	571	143.0	A[b]	>400
Valproate	261.0	578	2.2	A[a,b]	>1000
Phenobarbitone	3.4	71	20.9	A[a,b]	↑ 50
MK801(±)	0.1	0.3	3.0	A[a,b]	↑ 1

Dose response curves obtained from at least 5 doses (10 animals/dose).

IA = inactive, A = active at the 3 × ED50 dose.

[1]Oral ED50 MES test at 30 min (mice) and 60 min (rats) postdose.

[2]Oral TD50 inverted screen test (mice) at 30 min and gang plank escape test (rats) at 60 min post dose.

[3]TI = therapeutic index [TD50/ED50].

[4]ED50 (ip) for NMDA-induced convulsions (mice).

[5]Effects on development of kindled seizures [a] or inhibition of established kindled seizures [b].

[6]SMA = spontaneous motor activity measured in the 'Optovarimex', the values represent the lowest doses in which motor activity is influenced[5].

and FPL 12495AA in this system were 6.41 μM and 0.62 μM, respectively. The latter value is similar to cerebrospinal fluid concentrations of this metabolite after effective doses in rats.

The development of tolerance to remacemide was evaluated in rodents by administration of 3 × ED50 doses of the drug for five days. The ED50 for protection against MES was then compared to an untreated control group of animals. The ED50 in the treated animals was higher than controls suggesting the possibility of tolerance. In separate studies, hexobarbitone sleep time was

150

determined in control animals and in animals treated with ED50 or 3 × ED50 doses of remacemide for one week. On day 1, sleep time in the remacemide-treated animals was prolonged at both doses relative to controls, suggesting initial inhibition of hexobarbitone metabolism. However, no difference in sleep time was observed on Day 7 between the remacemide-treated animals and controls, suggesting induction of hexobarbital metabolism by virtue of the difference between day 1 and day 7 sleep time.

Preclinical toxicology

The median lethal single dose of remacemide in rats after oral administration was approximately 900 mg/kg and did not differ significantly between sexes. After intravenous administration, the median lethal dose was approximately 50 mg/kg in male and female animals.

Thirty-day toxicology studies in rats showed remacemide to be well tolerated at doses up to 100 mg/kg/day. Mean liver weights increased in a dose-related manner. This was considered to be an adaptive response to the increased metabolic load. Thirty-day toxicology studies in dogs revealed no gross or microscopic evidence of treatment-related abnormalities at doses up to 60 mg/kg. There was evidence of central nervous system stimulation, and seizures occurred in some animals in the 40–60 mg/kg dose range during the first week of dosing.

Ninety-day toxicology studies in rats showed remacemide to be well tolerated at doses up to 80 mg/kg/day. Liver weights were increased at doses of 80 mg/kg/day and above. Ninety-day toxicology studies in the dog revealed no gross histopathologic or laboratory findings. Elevations in terminal liver weight were seen in the highest dose group (60 mg/kg/day). Clinical signs (including seizures) were seen at doses of 30–60 mg/kg/day in a dose-related manner, but the severity and frequency were higher during the first week of dosing.

Remacemide had no effect on male fertility in mice at doses up to 100 mg/kg/day for 1 month. No embryotoxic or teratogenic effects were seen in rats at doses up to 100 mg/kg/day or in rabbits at doses of up to 75 mg/kg/day. The Ames test in 5 strains of Salmonella typhimurium was negative, with or without microsomal activation.

Neuropathologic studies in the brain of animals sacrificed after dosing with the NMDA-receptor antagonists, MK801, phencyclidine (PCP), ketamine and tiletamine have demonstrated vacuolization of some neurones in the posterior cingulate and retrosplenial neocortices[7]. This was an acute, reversible effect which reached a maximum between 4 and 12 hours and returned to normal by 24 hours. No vacuolization or other abnormal pathology was observed in brain sections taken from rats eight hours after subcutaneous administration of 80 mg/kg remacemide.

Metabolism, pharmacokinetics and drug interactions

After intravenous administration of a single dose of 15 mg/kg remacemide to male rats the drug was extensively distributed (Vd=3.3 L/kg) and rapidly eliminated ($t_{1/2}$=0.3 hr). After oral administration of a single dose of 80 mg/kg, the compound was rapidly absorbed (tmax=0.5 hr) and eliminated with an apparent half life of 3.7 hours. The bioavailability was approximately 12%. Renal clearance of the parent drug was less than 1% of the total clearance, suggesting extensive hepatic metabolism. Following a single p.o. dose of 15 mg/kg ^{14}C-remacemide to rats, seventy-six percent of the radiolabel was absorbed relative to intravenous administration of the same dose. The apparent elimination half life of total radioactivity was 6.4 hours. In combination with the data obtained from studies with cold drug, these observations suggest extensive first pass metabolism of remacemide in the rat. After intravenous administration of 10 mg/kg remacemide to dogs, the drug was extensively distributed (Vd=3.8 L/kg) and rapidly eliminated ($t_{1/2}$eq1.1 hr). After oral administration of 10 mg/kg in capsule form, the drug was rapidly absorbed (tmax=1.3 hr) and eliminated with an apparent half life of 1.2 hours. The bioavailability was approximately 42%. The elimination of unchanged remacemide in the urine was minimal.

After single oral doses of 10–400 mg remacemide in healthy volunteers, the drug was rapidly absorbed (tmax=1–2 hours) and eliminated with a mean half life of 4 hours, regardless of dose. Peak plasma concentrations after single doses of 150 mg were similar to those observed in rats one hour after ED50 doses of the drug. Concentrations of FPL 12495XX (the desglycinated metabolite) could not be detected in plasma at doses up to 300 mg. At higher doses the concentrations of this metabolite were low and variable. Single doses of remacemide up to 300 mg were well tolerated. Above this dose, lightheadedness and gastrointestinal upset were observed with increasing frequency.

After multiple oral dosing of remacemide in healthy volunteers (25 mg qid –150 mg qid for 5 days), the pharmacokinetic characteristics of the drug were consistent with those seen after single doses and the half life was similar (approximately 4 hours). Steady state plasma concentrations of the parent drug were achieved after 24 hours and remained constant over the remainder of the dosing period. Plasma concentrations of the desglycinated metabolite were quantitated at all dose levels, although they were close to the assay detection limits at the lowest dose. Steady state peak concentrations of this metabolite were achieved over 2-3 days and fluctuations during the 6 hour dose interval were very small. The apparent elimination half life of this metabolite was 12–24 hours and appeared to be independent of dose. The drug was generally well tolerated in the dose range studied. Lightheadedness and gastrointestinal upset were seen with increased frequency at the higher doses but the former appeared to resolve on continued administration.

The metabolic fate of remacemide after single doses has been partially determined in man. Approximately 25% of the administered dose has been accounted for in urine. This was present as the parent compound (FPL 12924XX), the desglycinated metabolite (FPL 12495XX), the benzyl- and phenyl-p-hydroxy metabolites of these two compounds (FPL 14464XX, FPL 14431XX, FPL 14430XX and FPL 14465XX), and the desglycinated N-hydroxide (FPL 15053XX). These compounds were present primarily as their glucuronide conjugates.

Preliminary evaluation of the pharmacokinetics of remacemide in epileptic patients on monotherapy with carbamazepine or phenytoin has shown that the steady state plasma concentrations of the parent compound and the desglycinated metabolite are lower than those seen in healthy volunteers at similar doses (25 mg to 150 mg qid). This has been attributed to the induction of oxidative hepatic drug metabolism pathways by these compounds. The drug was well tolerated by patients when administered for up to one month at these doses. Minimal systemic or central nervous system adverse experiences were reported but gastrointestinal symptoms were similar to those reported by healthy volunteers. These observations are consistent with pharmacokinetic observations in patients versus healthy volunteers.

Evaluation of the effects of remacemide on interictal EEG phenomena and on seizure frequency is currently being evaluated in medically refractory patients.

Acknowledgement

This work was conducted in the Departments of Chemistry, Biology, Pharmacy, Drug Safety Evaluation, Drug Metabolism, Analytical Sciences and Clinical Research and Development at Fisons' Divisional Research and Development facility in Rochester, New York, USA.

References

1   Porter, R.J., Cereghino, J.J., Gladding, G.D., Hessie, B.J., Kupferberg, H.J., Scoville, G. and White, B.G. (1984). Clev. Clinic Q., 51: 293–305.

2   Foster, A.C., Wong, E.H.F. (1987). Brit. J. Pharmacol, 91: 403–409.

3   Kupferberg, H.J. (1989). Epilepsia, 30 (Suppl. 1): S51–S56.

4   Swinyard, E.A., Raddhakrishnan, N., Goodman, L.S. (1962). J. Pharmacol. Exp. Ther. 138: 337–342.

5   Dourish, C.T. (1987). In: A.J. Greenshaw and C.T. Dourish (eds.), Experimental Psychopharmacology. Humana Press, New Jersey, pp. 153–211.

6   Macdonald, R.L. (1989). Epilepsia, 30 (Suppl. 1): S19–S28.

7   Olney, J.W., LaBruyere, J., Price, M.T. (1989). Science, 244: 1360–1362.

New Antiepileptic Drugs (Epilepsy Res. Suppl. 3)
Editors: F. Pisani, E. Perucca, G. Avanzini, A. Richens
© 1991 Elsevier Science Publishers B.V. (Biomedical Division)

CHAPTER 20

Stiripentol

Jean C. Vincent

Biocodex Research Laboratories, Montrouge, France

Introduction

Stiripentol, structurally unrelated to other antiepileptic drugs, was selected for study due to the demonstration of anticonvulsant properties in rodents and a low level of toxicity. A detailed knowledge of the peculiar pharmacokinetic profile of the drug allowed its rational use in complex partial and refractory absence seizures.

Chemistry

Stiripentol or 4,4-dimethyl-1-(3,4 methylenedioxyphenyl)-1-penten-3-ol is a white crystalline powder. It has a molecular weight of 234. Stiripentol is soluble in ethanol and acetone, moderately soluble in chloroform and insoluble in water. It is a racemic compound. A high performance liquid chromatography method is used to assay stiripentol in body fluids.

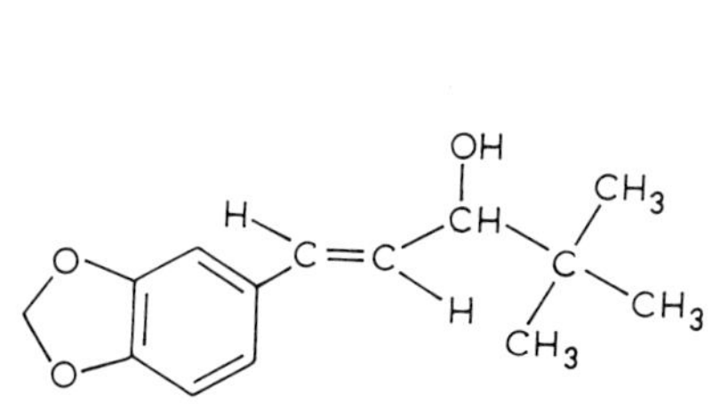

Fig. 1. Chemical structure of stiripentol.

Pharmacological properties

General pharmacological properties

Study of stiripentol actions on the central nervous system[1] showed that 200 mg/kg administered intraperitoneally in mice or rats increased the duration of barbiturate narcosis, decreased the amount of spontaneous movements, and rendered mice less aggressive. Rotarod-test and traction-test performances remained unchanged.

In the anesthetized dog receiving stiripentol, no significant changes were seen in the different hemodynamic parameters of the systemic circulation.

There was no development of gastric ulceration in the rat following administration of 200 mg/kg and there was no anti-inflammatory action in the paw carrageen test in the rat receiving stiripentol, 600 mg/kg p.o.

Anticonvulsant activity

Stiripentol produces a dose-related inhibition of electrically induced convulsions in rats. The ED_{50} was approximately 240 mg/kg i.p.[2].

Stiripentol antagonizes chemically induced convulsions. Administered after i.p. injection of pentylenetetrazol (PTZ) in mice it produced a dose-related protection against death. The ED_{50} of i.p. injected stiripentol for PTZ-induced convulsions was 200 mg/kg. Stiripentol given orally in rabbits 30 min after a first i.v. injection of PTZ reduced the formation of PTZ-specific paroxysmal waves.

Acute and chronic efficacy tests of stiripentol were conducted in alumina-gel rhesus monkeys[3]. In the acute study (n = 6), discrete serial seizures precipitated by 150 mg/kg of 4-deoxypyridoxine hydrochloride were antagonized by intravenous administration of stiripentol. In chronic studies stiripentol significantly reduced EEG interictal spike rates.

To test whether the combination of carbamazepine plus stiripentol is synergistic, an alumina-gel monkey model (n = 4) was used to compare polytherapy electroencephalographic (EEG) effects to those of CBZ monotherapy[4]. Relative to baseline, carbamazepine increased interictal EEG spikes by an average of 42%. Relative to carbamazepine monotherapy, the addition of stiripentol was associated with an average decrease in spike rate of 39%.

The carbamazepine-10,11-epoxide/carbamazepine ratio decreased from 0.29 to 0.06 when stiripentol was added and increased to 0.30 when stiripentol was discontinued.

The anti-absence effect of stiripentol was evaluated[5] in a strain of Wistar rats with a genetic petit mal-like epilepsy. Maximal suppression of SWD (spike and wave discharges) was respectively 68%, 83%, and 91% after i.p. injections of 125, 250 and 500 mg/kg.

The potential effectiveness of stiripentol against generalized epilepsy of the absence type was also evaluated in the intravenous PTZ infusion seizure model in the rat.

Significant elevation in PTZ threshold dose was observed at a single 300 mg/kg i.p. dose of stiripentol or at plasma levels exceeding 40 μg/ml. Maximal anticonvulsant response was reached at doses above 450 mg/kg (or plasma concentration > 120 μg/ml), along with the appearance of neurotoxicity. After subacute treatment, response

data were obtained at dosage levels of 150, 400 and 800 mg/kg with respective mean steady-state levels of 33.2 ± 7.8, 61.4 ± 20.7, and 116 ± 14 μg/ml. The degree of efficacy exhibited by stiripentol is comparable to that of the standard anti-absence drugs valproic acid and ethosuximide.

Mechanism of action

The neurochemical studies suggest that the anticonvulsant activity of stiripentol could be due to an increase of the cerebral GABA concentration by inhibition of its synaptosomal uptake and its metabolic turnover.

Clinical pharmacokinetics

Slow distribution of the drug influences the elimination curve making it multiphasic.

Plasma protein binding is extensive with a free fraction of 1%.

Stiripentol follows non linear pharmacokinetics. The non linear behaviour has several implications with respect to conditions of use: 1) therapeutic doses will generally be in the 2000–3000 mg/day range; 2) plasma level monitoring will be useful in dosage adjustments.

In epileptic patients, the clearance at steady state is threefold larger than in healthy subjects[8,9,10,11].

Stiripentol is the only known antiepileptic drug for which a complete mass balance has been established prior to its general availability for clinical use[12].

Activity and tolerability in epileptic patients

In a pilot efficacy study[13], stiripentol was substituted for other antiepileptic drugs and then associated with one of the baseline drugs. Of 21 patients, 3 under monotherapy with stiripentol remained seizure-free for periods of 27, 44 and 47 days. The other patients had to receive one of the baseline drugs and 12 of them experienced seizure reduction from 50 to 75% during at least 3 months of follow-up.

An open pilot study[14] was conducted in 7 patients with complex partial epilepsy incompletely controlled by CBZ and associated anticonvulsants. Carbamazepine dose was reduced by one third and stiripentol was progressively increased over 3 days to 1.5 mg/day. The mean seizure frequency significantly decreased from 12.9 ± 4.2/month at baseline to 3.9 ± 0.6/month after 4 months. Few side effects were observed. One patient reported transient irritability and 2 patients complained of somnolence. No changes in routine laboratory tests were observed.

A long-term, open trial[15] was performed in refractory patients with severe localization-related symptomatic epilepsy. Thirty one patients entered the study and 26 of them completed the 8-week evaluation period. During this period, 3 patients were seizure free, 7 had over 50% reduction in seizure frequency, 13 were slightly improved 6 showed an increase in seizure frequency.

In a phase 2 open study, stiripentol was tested in children with atypical absence seizures. Eight patients were enrolled and completed the study.

A target dose of stiripentol, according to weight, was instituted in each child and then adjusted to keep serum levels in the 10–18 µg/ml range. Other concomitant medications (carbamazepine and phenytoin) had their doses adjusted to keep serum levels similar to those observed during the baseline period. Valproic acid doses were not changed. Every patient experienced a substantial decrease (49 to 95%) in atypical absence seizures.

Few side effects were noticed. In one patient stiripentol had to be discontinued because of nausea, vomiting and weight loss.

Stiripentol was given as sole medication to 20 patients with refractory absence seizures, ranging in age from 13 to 46 years[16]. Five patients achieved a sustained seizure control and 6 other patients maintained an over 50% reduction in seizure frequency. Stiripentol was stopped in 4 patients because of lack of efficacy. Only 6 patients experienced mild side effects (gastric disturbances in 4 cases and insomnia in 2 cases).

A neuropsychological and efficacy evaluation of stiripentol[17] was completed in 11 patients with either a drug-resistant epilepsy or toxic effects ascribed to other anticonvulsants.

A trend towards an improvement ($p < 0.05$) in the performance of two tasks requiring sustained attention was noted. Previous side-effects (mainly drowsiness) decreased or disappeared in 7 of 9 patients who became more alert. Six of the nine uncontrolled patients experienced a decrease in seizure frequency of at least 50 percent.

Drug interactions

Stiripentol metabolism is induced by other antiepileptic drugs[13,14]. Stiripentol strongly inhibits the metabolism of other antiepileptic drugs[13,18,19,20,21,22]. Addition of stiripentol is associated with large increases in plasma levels of carbamazepine and phenytoin and, to a lower degree, phenobarbitone and primidone. These interactions are clinically relevant and require corresponding dosage reduction. Inhibition of carbamazepine metabolism by stiripentol is associated with a decreased carbamazepine-10,11-epoxide/carbamazepine ratio. This effect may partly explain the better tolerance of levels of carbamazepine above the usual therapeutic range.

References

1 Astoin, J., Marivain, A., Riveron, A., Crucifix, M., Laporte, M., Torrens, Y. (1978) Eur. J. Med. Chem. Ther. 13: 41–47.
2 Poisson, M., Huguet, F., Savattier, A., Bakri-Logeais,

156

F., Narcisse, G. (1984) Arzneim. Forsch./Drug. Res. 34: 199–204.

3 Lockard, J.S., Levy, R.H., Rhodes, P.H., Moore, D.F. (1985) Epilepsia 26: 704–712.

4 Lockard, J.S., Levy, R.H. (1988) Epilepsia 29: 476–481.

5 Micheletti, G., Vergnes, M., Lannes, B., Tor, J., Marescaux, C., Depaulis, A., Warter, J.M. (1988) Epilepsia 29: 709.

6 Lin, H.S., Levy, R.H. (1983) Epilepsia 24: 692–702.

7 Mesnil, M., Testa, B., Jenner, P. (1988) Biochem. Pharmacol. 37: 3619–3622.

8 Levy, R.H., Lin, H.S., Blehaut, H., Tor, J. (1983) J. Clin. Pharmacol. 23: 523–533.

9 Lin, H.S., Levy, R.H., Blehaut, H., Tor, J. (1984) In: Levy, R.H. (ed.), Metabolism of Antiepileptic Drugs. Raven Press, New York, pp. 199–207.

10 Levy, R.H., Loiseau, P., Guyot, M., Blehaut, H., Tor, J., Moreland, T.A. (1984) Epilepsia 25: 486–491.

11 Levy, R.H., Loiseau, P., Guyot, M., Blehaut, H., Tor, J., Moreland, T.A. (1984) Clin. Pharmacol. Ther. 36: 661–669.

12 Moreland, T.A., Astoin, J., Lepage, J., Tombret, F., Levy, R.H., Baillie, T.A. (1986) Drug. Metab. Disp. 14: 654–662.

13 Martinez-Lage, J.M., Loiseau, P., Levy, R.H., Gonzalez, I., Strube, E., Tor, J., Blehaut, H. (1984) Epilepsia 25: 673.

14 Rascol, O., Squalli, A., Montastruc, J.L., Garat, A., Houin, G., Lachau, S., Tor, J., Blehaut, H., Rascol, A. (1989) Clinical Neuropharmacology 12: 119–123.

15 Martinez-Lage, J.M., Levy, R.H., Gonzalez, I., Tor, J., Blehaut, H. (1987) In: Wolf, P., Dam, M., Janz, D., Dreifuss, F.E. (eds.), Advances in Epileptology, Vol. 16, Raven Press, New York, pp. 541–546.

16 Loiseau, P., Tor, J. (1987) Epilepsia 28: 579.

17 Loiseau, P., Strube, E., Tor, J., Levy, R.H., Dodrill, C. (1988) Revue Neurologique 144: 165–172.

18 Levy, R.H., Martinez-Lage, J.M., Tor, J., Blehaut, H., Gonzalez, I., Baindridge, B. (1985) Epilepsia 26: 544.

19 Levy, R.H., Martinez-Lage, J.M., Kerr, B.M., Viteri, C. (1987) 17th Epilepsy International Congress. Jerusalem.

20 Levy, R.H., Loiseau, P., Guyot, M., Acheampong, A., Tor, J., Rettenmeier, A.W. (1987) Epilepsia 28: 605.

21 Levy, R.H., Loiseau, P., Guyot, M., Acheampong, A., Tor, J. (1988) Epilepsia 29: 709.

22 Loiseau, P., Levy, R.H., Tor, J. (1989) In: Pitlick, W.H. (ed.), Antiepileptic Drug Interactions, Demos, New York.

New Antiepileptic Drugs (Epilepsy Res. Suppl. 3)
Editors: F. Pisani, E. Perucca, G. Avanzini, A. Richens
© 1991 Elsevier Science Publishers B.V. (Biomedical Division)

CHAPTER 21

Tiagabine[*]

M.W. Pierce[1], P.D. Suzdak[2], L.E. Gustavson[1], H.B. Mengel[2], J.F. McKelvy[1], T. Mant[3]

[1]*Abbott Laboratories, Abbott Park, IL, USA;* [2]*Novo Nordisk A/S, Bagsvaerd, Denmark;* [3]*Guy's Drug Research Unit, London, UK*

Introduction

γ-Aminobutyric acid (GABA)-mediated inhibitory systems have an important role in the regulation of seizure activity. Compounds that block the synthesis of GABA or inhibit the binding of GABA to its post-synaptic receptor are potent convulsant agents. Conversely, enhancement of GABA-mediated inhibition has an anticonvulsant effect. One mechanism for pharmacologically increasing GABA-mediated inhibition involves the blockade of GABA uptake into neurons or glia. Compounds such as nipecotic acid or guvacine preferentially block glial uptake of GABA and have an anticonvulsant action when injected intra-cerebroventricularly in mice[1]. A major problem with these compounds, however, has been their exclusion from the central nervous system by the blood-brain barrier. Tiagabine (NO-05-0328/A-70569 HCl), consisting of nipecotic acid joined by a linker to a lipophilic anchor, is a potent and specific inhibitor of GABA uptake into glial and neuronal elements in vitro[2]. Following administration by intraperitoneal injection or gastric lavage it has potent anticonvulsant activity in animal models. Tiagabine therefore represents a new pharmacological approach to the treatment of epilepsy.

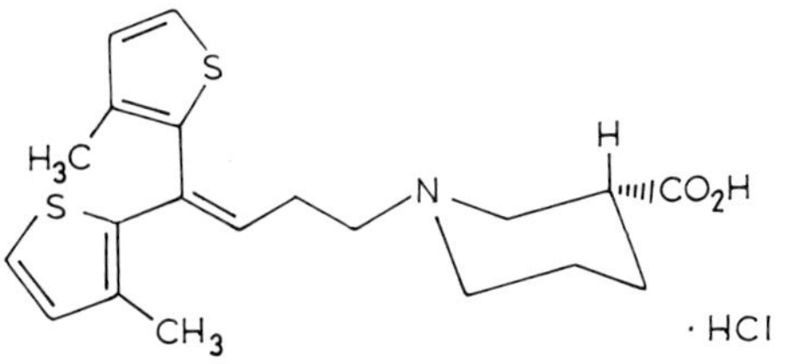

Chemistry

Tiagabine (R−)-N-(4,4-di-(3-methylthien-2-yl)but-3-enyl) nipecotic acid hydrochloride, MW 412.0, Figure 1) is a white to off-white, odourless, crystalline powder which melts with decomposition at 192°C. The UV spectrum has a maximum in water of 260 nm with an E_{mol} of 13000. There is no significant fluorescent property and it is non-hygroscopic. The enantiomeric purity of the compound is greater than 99% R(−) with a specific rotation in water at 20°C of −11. It has a solubility value of 3% in water but is practically insoluble in hexane. The pK_as of the −COOH and N+H groups are 3.3 and 9.4, respectively. The partition coefficient in octanol/water is 39.3 at pH 7.4.

Fig. 1. Chemical structure of tiagabine.

[*] Proposed INN name.

A sensitive and precise HPLC procedure has been developed for the determination of tiagabine (free base) concentrations in human plasma[3]. Isolation of tiagabine is achieved using solid-phase extraction on disposable C8 columns. Separation is performed on a C18 analytical column using a mobile phase containing the ion-pairing reagent sodium octansulfonate and the column effluent is monitored with colorimetric electro-chemical detection. The workup procedure recovers more than 95% of the drug from plasma. The precision of the method is good with coefficients of variation typically less than 5% for concentrations as low as 8 ng/mL. Coefficients of variation are higher at concentrations less than 8 ng/mL, but remain within acceptable limits (less than 17%) for concentrations as low as the limit of quantitation (2 ng/ml using 1 mL plasma samples). Tiagabine is stable in plasma, with no evidence of degradation after 23 hours at room temperature or two months at $-20°C$.

Pharmacological properties

Tiagabine inhibits the uptake of GABA into a rat forebrain synaptosomal membrane preparation with an IC_{50} value of 75 nM. In cell culture, it shows no clear specificity for neurons or glia, with IC_{50}s for inhibition of GABA uptake into neurons and glia of 446 and 182 nM, respectively[4,5]. Tiagabine does not inhibit the uptake of dopamine or noradrenaline (<20% inhibition at 100 μM). ^{3}H-tiagabine is neither a substrate for the uptake carrier process nor does it stimulate the release of GABA from neurones (concentrations less than 1 μM). It binds weakly to the central benzodiazepine receptor with an ED_{50} value of 4 μM, and enhances TBPS binding (a ligand which binds to the chloride channel associated with the GABA receptor) at concentrations greater than 3 μM. Tiagabine inhibits binding to histamine and 5-hydroxytryptamine $(HT)_{1B}$ receptors at concentrations 20- to 400-fold higher than those inhibiting GABA uptake. At 100 μM, tiagabine inhibits less than 20% of the binding of specific ligands to the dopamine D_1 and D_2, muscarinic, 5-HT_2, $beta_1$, $beta_2$, $alpha_1$, $GABA_A$, and gly-cine$_A$ receptors. ^{3}H-tiagabine showed saturable and reversible binding to rat brain membranes in the presence of $NaCl$[5]. The specific binding of ^{3}H-tiagabine was blocked by known inhibitors of ^{3}H-GABA, with an affinity constant of 18nM and a B_{max} of 669 pmole per gram rat forebrain tissue[5].

Tiagabine is a potent anticonvulsant in several rodent models (Table 1). It has a good oral duration of anticonvulsant action after either 1 hour (ED_{50} for inhibiting DMCM-induced convulsions = 3.4 mg/kg), 6 hours (ED_{50} = 18 mg/kg), or 17.5 hours (ED_{50} = 70 mg/kg).

In male Wistar rats tiagabine produced a dose-dependent protection against amygdaloid kindled seizures, with an ED_{50} value of 3 mg/kg i.p. Near complete protection was produced at a dose of 10 mg/kg i.p. Conversely, bicuculline-induced convulsions were only partially antagonized by tiagabine (approximately 40% protection at 8.0 mg/kg, i.p.); larger doses produced variable or no antagonistic effect. Maximal electroshock (MES)-induced convulsions were weakly antagonized by tiagabine (ED_{50}=40 mg/kg, i.p.); complete protection against MES-induced convulsions was obtained at 100.0 mg/kg, i.p.

Tiagabine produced protection against photically-induced myoclonic seizures in the photosensitive baboon. At doses of 0.1 and 0.25 mg/kg i.v. tiagabine produced a 50–65% protection which lasted for approximately 3 hours and was not associated with neurological side effects. A dose of 1 mg/kg i.v. produced a 70% protection and was associated with impaired motor coordination, diffuse tremor and slow abnormal movements of the limbs[6].

In rotarod, traction, and exploratory loco-motor tests for sedation, tiagabine exerted ac-

TABLE 1

ANTICONVULSANT PROPERTIES OF TIAGABINE AND REFERENCE COMPOUNDS

	Model/seizure type (ED$_{50}$; mg/kg i.p.)			
	Audiogenic tonic[a]	PTZ clonic[b]	DMCM clonic[b]	PTZ clonic[c]
Compound	DBA/2 mice	NMRI mice	NMRI mice	rats
Tiagabine	0.4	1.3	1	6
SKF 100330A	2.5	10	4	12
Carbamazepine	7	>200	15	>64
Clonazepam	0.004	0.035	0.9	0.56
Diazepam	0.17	0.7	2.5	3.5
Phenobarbitone	3.6	32	18	20
Phenytoin	11	>500	>100	>84
Valproic Acid	38	285	200	310

[a]Audiogenic seizures were induced in male DBA/Z mice by 111 dB 14 kHz sound stimulation.
[b]Seizures were chemically induced in male NMRI mice with 120 mg/kg s.c. pentylenetetrazol (PTZ) or in female mice with 15 mg/kg i.p. of 6,7-dimethoxy-4-ethyl-β-carboline-3 carboxylate (DMCM).
[c]Seizures were chemically induced in male Wistar rats with 120 mg/kg s.c. PTZ.

tivity at doses higher than those eliciting anticonvulsant activity. The ratios of anticonvulsant activity (ED$_{50}$) against DMCM seizures to sedative activity (ED$_{50}$) on either traction or the exploratory locomotor test in mice (i.p.) were 12.5 and 31.1, respectively. In rats, the ratio of anticonvulsant activity against PTZ-induced seizures to sedative effects in inhibition of tractions was 2.5. Animals given larger 'sedative' doses of tiagabine ($\geq$ 30 mg/kg i.p.) showed a syndrome consisting of a flat, rigid body posture, tremor, and tonic forelimb movements.

Clinical pharmacokinetics

The pharmacokinetics of tiagabine were examined in healthy subjects following single oral dose administration of 2, 8, 12, and 24 mg as part of a tolerance and safety study. The free base was rapidly absorbed in most subjects with maximum concentrations (C$_{max}$) occurring at 0.5 or 1.0 hours post-dose, but due to inter-individual variation C$_{max}$ occurred as late as 3 hours post-dose

in one subject. Mean peak plasma concentrations were 43.4, 150, 241, and 552 ng/mL for the 2, 8, 12, and 24 mg dose groups, respectively. The mean areas under the plasma concentration-time curves were 228, 727, 1450 and 2190 ng·hr/mL for the 2, 8, 12, and 24 mg doses, respectively. Dose-adjusted C$_{max}$ and areas under the curves were independent of the administered dose with overall means of 20.8 ng/mL/mg and 105 ng·hr/mL/mg, respectively. The apparent half-lives were quite variable, ranging from 4.5 to 13.4 hours, but also appeared to be independent of the administered dose. The overall harmonic mean half-life was 6.7 hours. Pharmacokinetic parameters obtained at day 1 and day 5 in a five day multiple dose trial were in accordance with the results of the single dose study. The dose levels tested were 2, 4, 6, 8, and 10 mg daily.

Tolerability in normal subjects

Tiagabine has been administered to humans in a double-blind, placebo-controlled, parallel

160

group study in which healthy male subjects were divided into four serial groups and administered single oral doses of 2, 8, 12, and 24 mg. There were no significant signs or symptoms in any subject following 2 and 8 mg tiagabine. At 12 mg, subjective effects included dizziness, light-headedness, poor concentration, slow response, and nausea (in one subject). The intensity of these events was mild to moderate; one subject experienced no symptoms. At 24 mg, three out of four subjects had moderate effects similar to those seen at 12 mg, but the fourth was disoriented and confused for a period of 6–8 hours. In all subjects, the symptoms were transient and self-limited. No clinically significant changes in vital signs, ECG, clinical chemistry, haematology, or urinalysis were noted.

In a five day multiple dose, double-blind, placebo-controlled, two-period cross-over study, doses of 2, 4, 6, 8, or 10 mg tiagabine daily were administered in ascending order to five serial groups of male volunteers. A number of mild and transient symptoms were reported during the study. Difficulty in concentrating, headache, and dizziness occurred more frequently after tiagabine than placebo and seemed to be dose related.

None of these effects were statistically significant when measured in a battery of psychometric tests. No clinically significant changes were observed in vital signs, ECG, haematology, or urinalysis. Minor, clinically insignificant elevations in transaminases were seen in a minority of subjects, but with no apparent relation to time or dose. In conclusion, tiagabine was generally well tolerated in doses up to 10 mg daily for five days.

References

1 Croucher, M.J., Meldrum, B.S., Krogsgaard-Larsen, P. (1983) Eur. J. Pharmacol. 89: 217–228.
2 Nielsen, E.B., Wolfbrandt, K.H., Anderson, K.E., Knutsen, L.J.S., Sonnenwald, U., Braestrup, C., (1987) Soc. Neurosci. Abst 13: 968.
3 Gustavson, L.E., Chu, S.-Y. (1989) Scientific report PPRd/89/223, Abbott Laboratories, Abbott Park, IL, U.S.A.
4 Braestrup, C., Nielsen, E.B., Wolfbrandt, K.H., Anderson, K.E., Knutsen, L.J.S., Sonnewald, U. (1987) In: Rand, M.J. and Raper, C. (eds.), Pharmacology, Elsevier, New York, pp. 125–128.
5 Braestrup, C., Nielsen, E.B., Sonnewald, U., Knutsen, L.J.S., Andersen, K.E., Jansen, J.A., Frederiksen, K., Andersen, P.H., Mortensen, A., Suzdak, P.D. (1990) J. Neurochem. 54: 639–647.
6 Meldrum, B. unpublished data.

New Antiepileptic Drugs (Epilepsy Res. Suppl. 3)
Editors: F. Pisani, E. Perucca, G. Avanzini, A. Richens
© 1991 Elsevier Science Publishers B.V. (Biomedical Division)

CHAPTER 22

Vigabatrin

John P. Mumford, Peter J. Lewis

Merrell Dow Research Institute, Winnersh, Berkshire, UK

Introduction

Vigabatrin (Sabril®) is a novel anticonvulsant drug and a structural analogue of the neurotransmitter GABA. It was first synthesised in 1974 as one of a series of enzyme-activated, irreversible inhibitors of GABA transaminase. It is readily absorbed after oral administration, producing a prolonged increase in CNS GABA concentration. The drug is effective in several animal models of epilepsy. Vigabatrin was first administered to man in 1979 and since that time, a large number of clinical studies have confirmed the efficacy of vigabatrin in severe refractory epileptic patients. Approximately half of the patients treated show a reduction in seizure frequency of more than 50%, including some 15% with a reduction of more than 75%. Acute and reversible side effects appeared to resemble those of other anticonvulsant drugs, the most common being drowsiness, fatigue, dizziness and behavioural changes. The clinical introduction of vigabatrin has been cautious, with the drug being available at present only for those patients whose epilepsy has proved refractory to other forms of therapy. Vigabatrin has been marketed in Great Britain since 1989.

Chemistry

Vigabatrin (γ-vinyl GABA; GVG; dl-4-aminohex-5-enoic acid) belongs to a group of enzyme inhibitors variously referred to as 'k$_{cat}$' inhibitors[1], and 'enzyme-activated irreversible inhibitors'[2]. Such compounds have latent reactive grouping, and are accepted as a substrate by the target enzyme, following which the enzyme is irreversibly inactivated by its own normal catalytic activity. Vigabatrin must be able to substitute for the natural substrate, and would there-

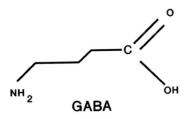

Fig. 1. Chemical structures of GABA and vigabatrin.

fore be expected to demonstrate stereospecificity in its action. Only the (S)-form is active, corresponding to observations in animal models of epilepsy that the (R)-form has no anticonvulsant activity.

Pharmacological properties

Role of GABAergic mechanisms in aetiology of epilepsy

GABA is the principal inhibitory transmitter in mammalian brain, acting at the $GABA_A$ receptor to increase membrane chloride conductance and thereby stabilise the resting membrane potential and at the $GABA_B$ receptor, which is found on presynaptic terminals and on postsynaptic membranes.

Enhancement of GABA-mediated inhibition

As GABAergic neurones exert an inhibitory effect, and as injection of compounds known to disrupt the synthesis or activity of GABA causes epileptiform convulsions, drugs that enhance GABA-mediated inhibition would be expected to have an anticonvulsant effect. GABA itself penetrates the blood-brain barrier only poorly[3]. Some GABA receptor agonists, e.g. muscimol, are anticonvulsant in several animal models, but may be pro-convulsant in others. Other agonists such as THIP have been demonstrated to possess a broad spectrum of anticonvulsant activity in mice, but clinical results have been disappointing[4].

Various compounds, including amino-oxyacetic acid and 1-cycloserine, can inhibit GABA transaminase, although the multiple biochemical actions of such agents meant that a simple correlation between anticonvulsant activity and elevation of brain GABA could not be established. The introduction of the irreversible GABA transaminase inhibitors, of which vigabatrin is one, finally confirmed the relationship between anti-convulsant action and GABA transaminase inhibition.

Effects of vigabatrin on GABAergic transmission

An important site of action of vigabatrin would appear to be in the midbrain, since seizure protection is correlated with GABA elevation in ventral midbrain following vigabatrin injection into ventral midbrain in animal models. Injections into the caudate, thalamus, superior coliculus and pontine regions are ineffective[5].

The effect of vigabatrin at clinically relevant concentrations on GABA transaminase and on release of endogenous GABA has been investigated in cultured astrocytes and neurones[6]. Vigabatrin was found to inhibit neuronal GABA transaminase preferentially.

To summarise, vigabatrin acts by irreversible inhibition of GABA transaminase, thereby allowing the accumulation of GABA in the CNS and subsequent suppression of epileptiform activity.

Biochemical effects in CNS

It has been shown[7] that at therapeutic doses in man vigabatrin produces dose-related increases in CSF concentrations of free and total GABA, and homocarnosine, biochemical changes which are consistent with inhibition of GABA transaminase activity. Relationship between the mean concentration of vigabatrin in blood, CSF, and of free GABA in the CSF in 11 patients following a single oral dose of 50 milligrams per kilogram body weight is shown in Figure 2. The concentration of vigabatrin in the blood and CSF is measured in nmols and that of free GABA in pmols. These changes are more marked when the dosing interval is decreased, corresponding to improved seizure control.

Other studies have attempted to correlate levels of GABAergic markers with seizure control. Twelve controls and 28 patients with refractory

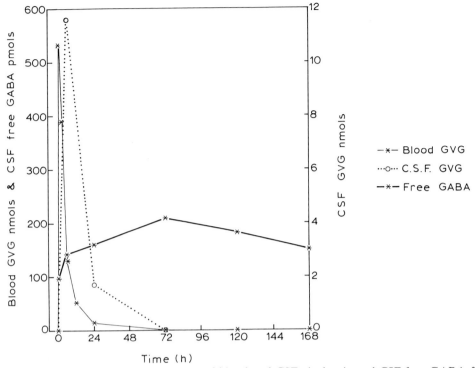

Fig. 2. Relationship between mean concentration of blood and CSF vigabatrin and CSF free GABA following a single oral dose of 50 mg/kg in 11 patients.

epilepsy were studied before and during 7 months of vigabatrin therapy. At baseline, total and free GABA in CSF of epileptic patients did not differ significantly from controls but during vigabatrin treatment, total GABA was 283%, free GABA 197% and homocarnosine 310% of baseline levels in the same patients[8].

Animal studies

Biochemistry. In mice, a single intraperitoneal dose of 1500 mg/kg vigabatrin caused a 5-fold increase in whole brain GABA within 4 hours[9]. This was paralleled by a sharp decrease in GABA transaminase activity, with recovery to only 60% of control values within 5 days.

In rats, injection of vigabatrin 100 mg/kg/day for 11 days produced marked changes in GABA, and homocarnosine. Similar, though less marked,

increases were seen following a single injection of the same dose[10].

Pharmacology. In mice injected with 1500 mg/kg vigabatrin intraperitoneally, the half life of the drug was estimated at approximately 16–17 hours. Vigabatrin was still detectable 3 days after administration[9].

In animals, vigabatrin has shown to provide protection against seizures induced by maximal electroshock, bicuculline and pentylenetetrazol when vigabatrin was injected directly into the midbrain of rats. Intravenous injection has been found to protect against bicuculline-induced myoclonic activity, strychnine-induced tonic seizures and isoniazid-induced generalized seizures, audiogenic seizures in mice, photic-induced seizures in the baboon and amygdala-kindled seizures in the rat[11–17].

Toxicology. A non progressive and reversible micro-vacuolization or intramyelinic oedema localised to certain white matter tracts of the CNS has been shown in rats and dogs following long term treatment[18,19]. Rats and dogs were given between 30 and 300 mg/kg/day vigabatrin for 12 to 24 months. Rats developed alopecia and convulsions and a fall in body weight with doses above 30 mg/kg/day. Post mortem examination showed intramyelinic oedema in rats given doses of vigabatrin above 30 mg/kg/day and dogs given 50 mg/kg/day or more. The lesion does not appear to be dose linked, and has never been reported in the peripheral nervous system. Demyelination has never been seen.

Pharmacokinetics

Following oral administration of 1–3 g vigabatrin orally to normal volunteers, the drug is rapidly absorbed, with peak plasma concentrations being reached within 2 hours[20]. The relative bioavailability of the drug was $92 \pm 11\%$. The plasma half-life is 5–7 hours, following linear kinetics, and the mean total plasma clearance is 1.75 ml/min/kg.

Elimination of vigabatrin occurs mainly by renal excretion, about 70% of the dose being recovered as unchanged parent compound in the urine within 24 hours of dosing. Non-renal clearance accounts for less than 30% in normal volunteers. Impairment in renal clearance of vigabatrin has been seen in elderly patients[21].

Vigabatrin is not bound to plasma proteins, and no metabolites have been identified in man. Food intake does not appear to have a clinically significant effect on vigabatrin absorption[22].

Clinical studies

Overview

In Europe, since 1984, seven double blind placebo controlled trials (six crossover and one parallel[23–29]) of up to 12 weeks' duration have confirmed the efficacy of vigabatrin in the treatment of chronic epilepsy in adults. Approximately 50% of patients show a reduction in seizure frequency of more than half, with 15% showing a reduction of better than 75%. In a few patients seizure frequency has increased on vigabatrin therapy. The most dramatic response has been in patients with partial seizures with or without secondary generalisation.

In an evaluation of over 1400 patients receiving vigabatrin as add-on therapy for drug resistant epilepsy in Europe, the percentage of those patients demonstrating a clinical benefit in that they had been continued on drug for over 6 months was greater for patients with partial type seizures than with generalised seizures[30]. Acute and reversible side effects resemble those of other anticonvulsant drugs, the most common being drowsiness, fatigue, dizziness and behavioural changes, with some patients also reporting weight gain.

The total daily dose of vigabatrin has varied from 1.5 to 3 g daily, usually given in divided doses, although once-daily dosing has also been reported as effective[31].

Neurophysiological and neuropathological studies

Somatosensory evoked potentials in dogs showed a significant slowing to the central transmission time 5 weeks after the initiation of vigabatrin therapy at a dose of 500 mg/kg/body weight[32]. These changes were associated with microvacuolation at 12 weeks. Both findings were fully reversible on stopping treatment.

In humans, no similar changes have been found. Hammond and Wilder[33] reported no change in EEG, visual evoked potentials, brainstem auditory potentials or cortical evoked potentials in 15 patients studied over 1 year's treatment with vigabatrin nor in 6 patients fol-

MAIN SEIZURE TYPE (n)

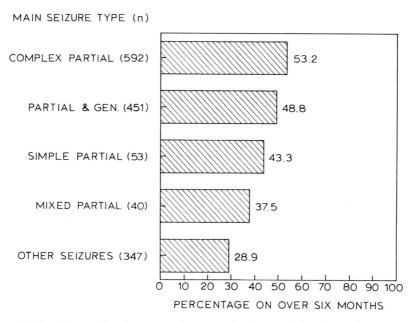

Fig. 3. Percentage of 1483 patients with refractory epilepsy continuing on vigabatrin – effect of the main seizure type.

lowed for 7 years. Kälviäinen and colleagues[34] and Liegeois-Chauvel and co-workers[35], reported no changes in visual evoked potentials or cortical evoked potentials in patients treated with vigabatrin for up to 18 months.

Chiron and co-workers[36] have used magnetic resonance imaging to investigate 16 children with epilepsy, before and after vigabatrin treatment. No abnormalities were found following an average of 11 months' treatment.

Butler[37] has reported no evidence of microvacuolation of myelin in 6 post-mortem and 17 operative specimens of human brain, following a mean duration of vigabatrin treatment of 27.8 months and 18.5 months respectively.

Psychometric and performance studies

Birbeck and co-workers[38] studied 27 patients who received 2–3 g vigabatrin daily for up to 1 year, and reported there appeared to be no significant mood changes related to vigabatrin. Gis-

selbrecht and Mumford[39] and McGuire and colleagues[40] reported that vigabatrin did not affect cognitive functions, and that there might even be some improvement in mental function.

Major outcome clinical trials of vigabatrin

Adults. In the multicentre study of Browne and co-workers[41] vigabatrin was given in a single blind design to 89 patients with complex partial seizures refractory to conventional therapy. The median number of seizures per month decreased from 11 to 5 after addition of vigabatrin, and 51% of patients showed a decrease of 50% or greater. Side effects were more frequent on initiation of therapy, and by 12 weeks only 13% of patients still reported them as troublesome. Sixty-six patients who showed a good response to vigabatrin were followed up for 6–54 months: only 17 patients dropped out due to break-through seizures and/or side effects, and no serious systemic or neurological toxicity was reported.

166

Chauvel and colleagues[42] evaluated the efficacy and tolerance of add-on vigabatrin therapy in 86 drug resistant epileptic patients aged 10–64 years. The diagnosis was focal epilepsy in 95% of patients. Vigabatrin was discontinued in three patients because of adverse reactions, and 25 patients withdrew from the study because of lack of efficacy. Of the 58 responders who continued therapy, 30 patients showed a reduction in seizure frequency of 75–100% and 14 patients 50–74%. There was no evidence of a decrease in efficacy with continued vigabatrin therapy.

Tartara and co-workers[43] reported similar findings in a 3-year prospective study of twenty-five patients. Eight patients withdrew because of side effects, lack of efficacy, or other reasons and seventeen patients continued on drug for a period ranging from 25–57 months. Response to vigabatrin was maintained over time, and there was no evidence of neurotoxicity.

Sivenius and colleagues[44] have reported a double blind dose reduction study of vigabatrin in seventy five patients with drug resistant complex partial epilepsy. Of the 53 responders, 28 continued with 3 g/day and 25 had their dose reduced to 1.5 g/day under double blind conditions. Patients continuing on 1.5 g/day still had less seizures than during baseline and those on 3 g/day maintained the same efficacy.

Children. Livingstone and co-workers[45] reported on the efficacy of vigabatrin in 135 children, aged 2 months to 12 years. Seizure types were mainly partial (42%) or generalized (29%). Vigabatrin was added to existing therapy with a final mean dose of 87 mg/kg/day. A better than 50% reduction in seizure frequency was observed in 38% of patients. A better response was also noted in partial seizures. Agitation and insomnia were reported in 8.8% and somnolence in 6% with ataxia (2.2%), nausea (2.2%) and increased appetite (1%) also observed. No side effects were

observed in 79% of patients. Luna and colleagues[46] reported a single blind placebo controlled study of vigabatrin in 61 children (12 months to 19 years) with refractory epilepsy. Doses ranged from 50–150 mg/kg/day. Twenty three patients (38%) showed a reduction of more than 50% in seizure frequency though twelve patients (20%) showed an increase of seizures. These workers found that vigabatrin appeared particularly effective in partial epilepsy, but non progressive myoclonic epilepsy tended to be aggravated. Chiron and colleagues[47] have studied the effects of vigabatrin in 45 children with infantile spasms. Response appeared to be better in children with symptomatic infantile spasms (11 of 16 children responding) than those with cryptogenic infantile spasms.

Tolerance and side effects

The most frequently reported side effects of vigabatrin therapy have been drowsiness, ataxia and weight gain and these have usually been transient.

Remy and Beaumont[49] reported a multicentre study of the long term safety of vigabatrin in 254 patients in 23 different clinics in 8 European countries. 75% of patients reported no side effects at all. CNS symptoms seemed to predominate, with 55 out of 254 patients reporting sedation. Irritation, aggression and memory problems were reported by 18.5%. Weight gain was reported by 6.7% of patients, and the incidence of other effects was below 3%. Side effects were usually mild (58.2%) or moderate (31.3%): only 10.5% were classed as severe. Four patients had vigabatrin therapy discontinued because of adverse effects.

Less than 3% of examinations performed showed any evidence of neurological changes on vigabatrin therapy. Laboratory safety tests showed a small (4%) decrease in haemoglobin and haematocrit, although individual values remained within normal limits. There was also some alteration in liver transaminases (10–12% decrease in SGOT and 30–50% decrease in SGPT) which is thought to be an artefactual inhibition of these transaminases and of no biological significance.

Dosage

Vigabatrin is usually given in doses of 1–3 g/day, in divided doses, although a recent study[31] conducted over a period of 9–34 months has reported that once-daily dosing is also effective in the treatment of epilepsy.

Drug interactions

Vigabatrin is usually administered as add-on therapy, patients are usually taking other anticonvulsant drugs. The only drug interaction reported with vigabatrin is a fall in plasma phenytoin concentrations of approximately 23%[48]. The mechanism of this interaction is uncertain, but it does not involve protein binding or a change in phenytoin metabolism. The authors have postulated that it may result from a delayed effect which could not be observed in their study, and further studies are required.

Acknowledgements

To the numerous clinical investigators throughout Europe who have assisted so ably in the evaluation of vigabatrin and to Dr Susan Boobis for her help in the preparation of this manuscript.

References

1 Rando, R.R. (1974) Science 185: 320–324.
2 Seiler, N., Jung, M.J., Koch-Weiser, J. (eds.) (1978) In: Enzyme-activated Irreversible Inhibitors, Elsevier, Amsterdam.
3 Meldrum, B.S. (199) Brit. J. Clin. Pharmacol. 27: 3S – 11S.
4 Gram, L. (1988) Thesis, University of Copenhagen, Denmark.
5 Gale, K. (189) Epilepsia Suppl. 3: S1– S11.
6 Gram, L., Larsson, D.M., Johnsen, A., Schousboe, A. (1989) Br. J. Clin. Pharmacol. 27: 13S –17S.
7 Ben-Menachem, E., Persson, L.I., Schechter, P.J. (1989) Br. J. Clin. Pharmacol. 27: 79S –85S.
8 Pitkanen, A., Matilainen, R., Tuuttainen, T., Riekkinen, P. (1988) J. Neurol. Sci. 88: 83–93.
9 Jung, M.J. Lippert, B. Metcalf, B.W., Bohlen, F., Schechter, P.J. (1977) J. Neurochem. 29: 797–802.
10 Perry, T.L., Kish, S.J., Hansen, S. (1979) J. Neurochem. 32: 1641–1645.
11 Gale, K. (1986) Adv. Neurol. 44: 343–364.
12 Kendall, D.A., Fox, A., Enna, S.J. (1981) Neuropharmacology 2: 177–183.
13 Schechter, P.J., Tranier, Y., Grove, J. (1979) In: GABA-biochemistry and CNS Functions, Mandel, P., De Freudis, F.V. (eds.), Plenum Press, New York 43–57.
14 Schechter, P.J., Tranier, Y., Jung, M.J., Bohlen, P. (1977) Eur. J. Pharmacol. 45: 319–328.
15 Meldrum, B., Horton, R. (1978) Psychopharmacology (Berlin) 59: 47–50.
16 Piredda, S., Lim, C.R., Gale, K. (1985) Life Sci. 36: 1295–1298.
17 Stevens, J.R., Phillips, I., De Beaurefaire, R. (1988) Epilepsia 29: 494–411.
18 Butler, W.H., Ford, G.P., Newberne, J.W. (1987) Toxicol. Pathol. 15: 143–148.
19 Hauw, J.J., Trottier, S.M., Boutry, J.M., Sun, P., Sazdovitch, V., Duyckaerts, C. (1988) Br. J. Clin. Pract. 42 S61: 10–13.
20 Schechter, P.J. (1989) Br. J. Clin. Pharmacol. 27.S1: 19S –22S.
21 Haegele, K.D., Huebert, N.D., Egel, M., Tell, G.P., Schechter, P.J. (1988) Clin. Pharmacol. Ther. 44: 558–565.
22 Frisk-Holmberg, M., Kerth, P., Meyer, F. (1989) Br J Clin Pharmacol 27.S1: 23S –25S.
23 Rimmer, E.M., Richens, A. (1984) Lancet 1: 189–190.
24 Gram, L., Klosterkov, P., Dam, M. (1985) Ann. Neurol. 17: 262–266.
25 Loiseau, P., Hardenberg, J.P., Pestre, M., Guyot, M., Schechter, P.J., Tell, G.P. (1986) Epilepsia 27: 115–120.
26 Tartara, A., Manni, R., Galimberti, C.A., Hardenberg, J., Orwin, J., Perucca, E. (1986) Epilepsia 27: 717–723.
27 Tassinari, C.A., Michelucci, R., Ambrosetto, G., Salvi, F. (1987) Arch. Neurol. 44: 907–910.
28 Remy, C., Favel, P., Tell, G., Schechter, P.J. (1986) Boll. Lega Ital. Epilessia 54/55: 241–243.

29 Reynolds, E.H., Ring, H., Heller, A. (1988) Br. J. Clin. Pract. 42.S61: 33.

30 Mumford, J.P., Lewis, P.J. (1990) Merrell Dow data on file.

31 Ben-Menachem, E., Persson, L., Mumford, J.P. (1990) Epilepsy Res. 5: 240–246.

32 Arrezzo, J.C., Schroeder, C.E., Litwak, M.S., Stewart, D.L. (1989) Br. J. Clin. Pharmacol. 27.S1: 53S –60S.

33 Hammond, E.J., Wilder, B.J. (1989) In: Abstracts of the 18th International Epilepsy Congress: 37.

34 Kälviäinen, R., Partanen, J., Saksa, M., Sivenius, J., Riekkinen, P.J. (1989) In: Abstracts of the 18th International Epilepsy Congress: 90.

35 Liegeois-Chauvel, C., Marquis, F., Gisselbrecht, D., Fantieri, R., Beaumont, D., Chauvel, P. (1989) Br. J. Clin. Pharmacol. 27.S1: 69S –72S.

36 Chiron, C., Desguerre, Mondragon, S., Luna, D., Beaumont, D., Dulac, O. (1989) In: Abstracts of the 18th International Epilepsy Congress: 140.

37 Butler, W.H. (1990) J. Child Neurology (accepted).

38 Birbeck, K.A., Ossetin, J., Ring, H.A. (1989) In: Abstracts of the 18th International Epilepsy Congress: 78.

39 Gisselbrecht, D., Mumford, J.P. (1989) In: Abstracts of the 18th International Epilepsy Congress: 149.

40 McGuire, A.M., Duncan, J.S., Trimble, M.R. (1989) In: Abstracts of the 18th International Epilepsy Congress: 31.

41 Browne, T.R., Mattson, R.H., Penry, J.K. (1989) Br. J. Clin. Pharmacol. 27.S1: 95S –100S.

42 Chauvel, P., Vignal, J.F., Trottier, S., Beaumont, D. (1989) In: Abstracts of the 18th International Epilepsy Congress: 140.

43 Tartara, A., Manni, R., Galimberti, C.A., Zucca, C., Iudice, A., Perucca, E. (1989) In Abstracts of the 18th International Epilepsy Congress: 81.

44 Sivenius, J., Ylinen, A., Matilainen, R., Murros, K., Riekkinen, P.J. (1989) In: Abstracts of the 18th International Epilepsy Congress: 90.

45 Livingston, J.H., Beaumont, D., Arzimanoglou, A., Aicardi, J. (1989) Br. J. Clin. Pharmacol. 27.S1: 109S – 117S.

46 Luna, D., Dulac, O., Pajot, N., Beaumont, D. (1989) Epilepsia 30: 430–437.

47 Chiron, C., Dulac, O., Luna, D. (1990) Lancet 335: 363–364.

48 Rimmer, E.M., Richens, A. (1989) Br. J. Clin. Pharmacol. 27.S1: 27S –33S.

49 Remy, C., Beaumont, D. (1989) Br. J. Clin. Pharmacol. 27.S1: 125S –129S.

New Antiepileptic Drugs (Epilepsy Res. Suppl. 3)
Editors: F. Pisani, E. Perucca, G. Avanzini, A. Richens

CHAPTER 23

Zonisamide

M. Seino[1], H. Miyazaki[2], T. Ito[2]

[1]*National Epilepsy Center, Shizuoka Higashi Hospital, Shizuoka;* [2]*Research Laboratories, Dainippon Pharmaceutical Co. Ltd.,*
Suita, Osaka, Japan

Introduction

Zonisamide (AD-810, CI-912) is a novel anti-epileptic drug developed in Japan by Dainippon Pharmaceutical Co., Ltd. through studies on both the structure-anticonvulsant activity relationship of 3-substituted benzisoxazole derivatives[1] and the pharmacodynamics of the drug in animal models[2,3,4]. The drug has been proven effective in the treatment of patients[5,6,7] with either simple/complex partial seizures and secondarily generalized tonic-clonic seizures, or generalized tonic-clonic seizures, tonic seizures, atypical absence seizures and a combination of these. Zonisamide has been marketed in Japan under the name 'Excegran' in 1989.

Chemistry

Zonisamide is a fine, white crystalline solid or powder with pKa=9.7 and very slight solubility in water (0.8 mg/ml). The compound is stable in various mammalian sera and at the 1–7 pH range in aqueous solutions for at least 5 days.

The drug concentrations in human or animal plasma can be measured by HPLC[8], but more conveniently by an enzyme immunoassay method[9] which is commercially available in a kit form ('MARKIT A Excegran', Dainippon Phar-maceutical Co., Ltd.).

Chemical name: 1,2-benzisoxazole-3-methane-sulfonamide

Empirical formula: $C_8H_8N_2O_3S$

Molecular weight: 212.2

Melting point: 164–168°C

Chemical structure:

Pharmacological properties

Convulsive manifestations

The anticonvulsant effects of zonisamide are characterized by the following 3 aspects. *i) The anticonvulsant profile:* Like phenytoin and carbamazepine, zonisamide suppresses maximal electroshock seizures (MES) without affecting the seizures induced by pentetrazol in mice[2] (Table 1). *ii) The wide therapeutic range:* ZNS exerts anti-MES effects in mice, rats, rabbits and dogs at minimal plasma concentrations of approximately 10 μg/ml and neurological side effects at concentrations over 70 μg/ml[10]. The therapeutic range of zonisamide is much wider than that of phenytoin or carbamazepine, although the protective index (neurotoxic dose NTD_{50}/anti-MES dose ED_{50}) of zonisamide is comparable to that of the latter two drugs (Table 1). *iii) No tolerance*

170

TABLE 1

ANTICONVULSANT AND NEUROTOXIC DOSES OF FOUR ANTIEPILEPTIC DRUGS IN MICE

Items	MES ED_{50} (mg/kg, p.o.)	s.c.Met ED_{50} (mg/kg, p.o.)	Neurotoxicity NTD_{50} (mg/kg, p.o.)
Zonisamide	19.6	> 500	228
Phenytoin	7.9	> 500	72
Carbamazepine	13.3	> 500	141
Valproate	591.0	316.0	1068

s.c.Met: pentetrazole 85 mg/kg, s.c.

to anti-MES effects: The anti-MES effect of zonisamide remained unchanged after repeated oral administration in rats, while that of phenytoin or carbamazepine was greatly reduced. Neither metabolically- nor pharmacodynamically-mediated tolerance is observed to the anti-MES effect of zonisamide[11].

Electroencephalographic (EEG) seizures

Based on the anti-MES effects, zonisamide, like phenytoin and carbamazepine, is suggested to exert its anticonvulsant effects by inhibiting the spread of seizures as evidenced by the epileptic EEG discharge. Namely, zonisamide restricts the spread of cortical focal discharges evoked in cats by electrical stimulation, and prevents the propagation from cortex to subcortical structures of discharges evoked by cortical freezing in the animals[3].

In addition, zonisamide suppresses the spiking activity induced by cortical freezing in cats and the spikes occurring interictally between generalized convulsive seizures induced by cortical application of tungstic acid gel in rats, suggesting a suppression of focal epileptogenic activity[3]. Such suppression is not observed with phenytoin and carbamazepine.

Zonisamide suppresses focal cortical seizure discharges in cats as described above, but is almost without effect on thalamic and hippocampal afterdischarges induced by electrical stimulation

in the animals[3], suggesting a selective drug effect on the cortex. However, zonisamide suppresses subcortically evoked seizures in animals with increased seizure susceptibility. Namely, zonisamide suppresses afterdischarges in cortical and hippocampal kindled rats[11] and generalized discharges evoked by thalamic application of tungstic acid gel in cats[4]. In addition, oral administration of zonisamide for several days abolishes both afterdischarges and clinical convulsions in amygdaloid overkindled cats[12].

From these results, a blockade of the spread or propagation of seizure discharges and a suppression of epileptogenic focus activity seem to be involved in the anticonvulsant action of zonisamide. The detailed mechanism of the action of zonisamide remains obscure at the present time, although some hypotheses have been made[4,11]. Additional pharmacological data are provided by the papers listed in the references[13,14].

Clinical pharmacokinetics

After oral administration of zonisamide to healthy volunteers, plasma levels reached rapidly a maximum concentration (Cmax) of 2–3 µg/ml at the dose of 200 mg[11,15,16], of around 5 µg/ml at 400 mg[11,16] and of 12–13 µg/ml at 800 mg[11,16]. In these studies the time of peak (Tmax) ranged between 2.4 and 6.0 hr. Elimination half lives were practically independent of the doses and ranged from 49.7 to 68.2 hr (mean 60 hr).

In patients on treatment with some conventional antiepileptic drugs, the half-life of zonisamide after a single oral dose of 400 mg was 28.4 hr[17]. The shorter zonisamide half-life was also reported in patients on monotherapy with carbamazepine or phenytoin[18]. Steady state plasma or serum levels (Css) of zonisamide in patients were higher[17,19] than those predicted from the single dose administration parameters, probably due to nonlinear kinetics[20]. For practical purposes, Css can be empirically regarded to be linearly proportional to dose within the therapeutic dose range (mg/kg/day) in adult[21] and pediatric[22] patients.

Like other sulphonamide derivatives, zonisamide is highly concentrated in erythrocytes[16], because of its binding to carbonic anhydrase and other red cell protein component(s) with higher affinities than extracellular serum albumin[23]. It is therefore recommended to measure the plasma or serum level of zonisamide for drug monitoring rather than the whole blood level which exhibits a complex relationship with the plasma concentration.

In man, zonisamide is mainly excreted in urine as unchanged drug, acetylated and glucuronide metabolites[16]. The glucuronide metabolite is produced by cleavage of the N-O bond of isoxazole ring moiety.

Activity and tolerability in epileptic patients

Clinical studies in Japan
Double-blind controlled study. Seino et al.[6] conducted a multicenter double-blind comparison of zonisamide and carbamazepine in patients with epilepsy uncontrolled with one to three antiepileptic drugs or not previously treated. The patients (n=123) were treated for 16 weeks with either zonisamide or carbamazepine at an average daily dose of 330 mg or 600 mg respectively at week 16. One hundred and eight patients were evaluated for efficacy and 116 for safety. The average frequency of simple/complex partial seizures per 4 weeks decreased from 14.9 to 3.4 in the zonisamide group, and from 13.3 to 4.4 in the carbamazepine group. A similar reduction (i.e. from 1.8 to 0.6 in the zonisamide group, and from 2.1 to 0.7 in the carbamazepine group) in seizures per 4 weeks was also found for secondarily generalized tonic-clonic seizures. The percentage of patients showing a $\geqslant 50\%$ reduction in seizure frequency versus baseline was 82% in the zonisamide group and 71% in the carbamazepine group at week 16. The overall improvement rate was 66% in the zonisamide group, and 65% in the carbamazepine group (Table 2). Although there was no statistically significant difference between the two groups in the incidence of both subjective and objective side effects (52% in the zonisamide group, and 57% in the carbamazepine group), there was a significantly higher incidence of anorexia in the zonisamide group, and of ataxia in the

TABLE 2

OVERALL IMPROVEMENT (ZONISAMIDE VS CARBAMAZEPINE)

Drug	Markedly improved	Improved	Slightly improved	Unchanged
Zonisamide N=56	17 (30%)	20 (36%)	4	15
Carbamazepine N=52	19 (37%)	15 (29%)	4	14

TABLE 3

SAFETY (ZONISAMIDE VS CARBAMAZEPINE)

Drug	Safe	Nearly safe	Uncertainly safe	Not safe
Zonisamide N=58	30 (52%)	18 (31%)	10	0
Carbamazepine N=58	24 (41%)	14 (24%)	17	3

carbamazepine group. The safety rating for the zonisamide and the carbamazepine group was 83% and 66% respectively, a statistically significant difference (p=0.08) (Table 3).

Other controlled studies. Oguni et al.[7] conducted an efficacy and safety evaluation of zonisamide and valproic acid in pediatric patients (n=32) with convulsive or non-convulsive generalized seizures having 4 or more seizures per month. The children were treated with one to three antiepileptic drugs or had received no medication. The zonisamide group and the valproic acid group were treated for 8 weeks at average daily doses of 7.3 mg/kg and 27.6 mg/kg respectively at week 8. Although there was no significant difference between the two groups in the outcome of clinical seizures, the percentage of patients showing a ≥50% reduction in seizures compared to baseline was 77% in the zonisamide group and 53% in the valproic acid group at week 8. The overall improvement rating was 50% in the zonisamide group and 44% in the valproic

acid group (Table 4). Like valproic acid, zonisamide was found to be useful in controlling generalized seizures such as tonic-clonic seizures, tonic seizures and atypical absences.

Open studies. Clinical data[5,6,7] on the efficacy and safety of ZNS (100-mg tablets and powder) were evaluated at 56 sites in a total of 1,008 patients with various epilepsies (605 cases, ≥16 years; 292, 7–15 years; 99, 2–6 years; and 12, ≤1 year). These studies included patients with partial (712 cases), generalized (163 cases), and mixed (132 cases) seizures. Despite treatment with an average of 2.8 conventional antiepileptic drugs, 77% of the patients had daily to weekly seizures. The duration of treatment at final evaluation was 253 days on average and 271 patients were treated for one year or longer. The overall improvement rate was 56%. The improvement rates in the add-on therapy were 61% in simple partial seizures, 52% in complex partial seizures, 61% in secondarily generalized tonic-clonic seizures, 53% in generalized tonic-clonic seizures, 35% in gen-

TABLE 4

OVERALL IMPROVEMENT (ZONISAMIDE VS VALPROIC ACID)

Drug	Markedly improved	Improved	Slightly improved	Unchanged	Slightly worse
Zonisamide N=16	5 (31%)	3 (19%)	4	4	0
Valproic acid N=16	5 (31%)	2 (13%)	2	5	2

eralized tonic seizures, 73% in atypical absence seizures, 30% in atonic seizures, 29% in myoclonic seizures, and 54% in the mixed generalized seizures. In monotherapy with zonisamide at the time of final evaluation, the improvement rates were 93% in simple partial seizures, 93% in complex partial seizures, 94% in secondarily generalized tonic-clonic seizures, 78% in generalized tonic-clonic seizures, and 43% in generalized tonic seizures. Side effects occurred in 517 cases (51%), of whom 185 (18%) discontinued the trial. Adverse effects consisted of drowsiness (24%), ataxia (13%), loss of appetite (11%), gastrointestinal symptoms (7%), loss or decrease of spontaneity (6%), and slowing of mental activity (5%). The side effects in the patients given zonisamide alone were drowsiness (9%), loss of appetite (7%), gastrointestinal signs (7%), loss or decrease of spontaneity (6%), headache (6%), skin rash/itching (6%) and loss of weight (6%). Seventeen patients (2%) discontinued the trial because of an increase in γ-GTP, ALP, GOT/GPT, or leukopenia.

Clinical studies in the United States and Europe. Loiseau et al.[24] performed a multicenter placebo-controlled double-blind study of zonisamide in epileptic patients (n=139) with refractory partial seizures (including secondarily generalized tonic-clonic seizures) in Europe. Their overall evaluation in 12 week treatment showed an improvement rate of 62% in the zonisamide group and 19% in the placebo group (p<0.05). The side effects related to zonisamide affected mainly the central neurvous system or the gastrointestinal tract, but were not so severe to require discontinuation of zonisamide therapy. Sackellares et al.[17] found a 75–100% reduction in seizure frequency in 4 of 10 patients with refractory seizures. Wilensky et al.[19] undertook a crossover study of phenytoin, carbamazepine and zonisamide in 8 patients with uncontrolled partial

seizures, and found that zonisamide was definitely effective in 5 patients. Two of these patients responded better to zonisamide than to phenytoin or carbamazepine. With regard to side effects, Berent et al.[25] reported that zonisamide given concurrently with 2 or 3 conventional antiepileptic drugs affected specific cognitive functions such as acquisition and consolidation of recent information concerning verbal items. Henry et al.[26] reported that zonisamide was effective in two patients with progressive myoclonus epilepsy of the Unverricht-Lundborg type.

In the U.S. and Europe, urinary calculi have been found in some patients treated with zonisamide and further clinical trials of the drug have been suspended. In Japan, however, only 2 out of 1008 patients were found to have urinary lithiasis in the open-label studies described in this chapter. Both patients experienced spontaneous elimination of the calculi. A history of urolithiasis/nephrolithiasis was found in either the mother or the father of each of these two affected patients.

Drug interactions

In vitro, no serum protein binding interactions have been observed between zonisamide and sulthiame, phenytoin or phenobarbitone[5]. In patients receiving maintenance treatment with carbamazepine and phenytoin (or a combination of these drugs with phenobarbitone, primidone or valproic acid), plasma elimination half-lives of zonisamide are reported to be shorter than in normal subjects[17,18]. Slight changes in steady-state plasma levels of zonisamide have been reported as a result of concomitant administration of carbamazepine[21] and valproic acid[22].

References

1 Uno, H., Kurokawa, M., Masuda, Y., Nishimura, H.

174

(1979) J. Med. Chem. 22: 180–183.

2 Masuda, Y., Karasawa, T., Shiraishi, Y., Hori, M., Yoshida, K., Shimizu, M. (1980) Arzneim. Forsch. 30: 477–483.

3 Ito, T., Hori, M., Masuda, Y., Yoshida, K., Shimizu, M. (1980) Arzneim. Forsch. 30: 603–609.

4 Ito, T., Hori, M., Kadokawa, T. (196) Epilepsia 27: 367–374.

5 Ono, T., Yagi, K. Seino, M. (1988) Clinical Psychiatry 30: 471–482 (in Japanese).

6 Seino, M., Ohkuma, T., Miyasaka, M., Manaka, S., Takahashi, R., Murasaki, M., Sakuma, A. (1988) J. Clin. Exptl Med. 144: 275–291 (in Japanese).

7 Oguni, H., Hayashi, K., Fukuyama, Y., Iinuma, K., Seki, T., Seino, M., Watanabe, K., Mimaki, T., Ootahara, S., Kurokawa, T., Kuriya, N. (1988) Jap. J. Pediatrics 41: 439–450 (in Japanese).

8 Juergens, U. (1987) J. Chromatogr. 385: 233–240.

9 Kaibe, K., Nishimura, S., Ishii, H., Sunahara, N., Naruto, S., Kurooka, S. (1990) Clin. Chem. 36: 24–27.

10 Masuda, Y., Utsui, Y., Shiraishi, Y., Karasawa, T., Yoshida, K., Shimizu, M. (1979) Epilepsia 20: 623–633.

11 Taylor, C.P. McLean, J.R. Bockbrader, H.N. Buchanan, R.A., Karasawa, T., Miyazaki, M., Rock, D.M., Takemoto, Y., Uno, H., Walker, R. (1986) In: Meldrum, B.S., Porter, R.J. (eds.), New Anticonvulsant Drugs. John Libbey & Co., Ltd., London, pp. 277–294.

12 Kakegawa, N. (1986) Psychiatria et Neurologia Japonica 88: 81–98 (in Japanese).

13 Hori, M., Ito, T., Oka, M., Noda, Y. Matsuno, Y., Furukawa, K., Ochi, Y., Karasawa, T., Kadokawa, T. (1987) Arzneim. Forsch. 37: 1124–1130.

14 Nakatsuji, K., Matsuno, Y., Nakamura, N., Fujitani, B., Ito, T., Kadokawa, T. (1987) Arzneim. Forsch. 37: 1131–1136.

15 Matsumoto, K., Miyazaki, H., Fujii, T., Kagemoto, A., Maeda, T., Hashimoto, M. (1983) Arzneim. Forsch. 33: 961–968.

16 Ito, T., Yamaguchi, T., Miyazaki, H., Sekine, Y., Shimizu, M., Ishida, S., Yagi, K., Kakegawa, N., Seino, M., Wada, T. (1982) Arzneim. Forsch. 32: 1581–1586.

17 Sackellares, J.C., Donofrio, P.D., Wagner, J.G., Abou-Khalil, B., Berent, S., Aasved-Hoyt, K. (1985) Epilepsia 26: 206–211.

18 Ojemann, L.M., Shastri, R.A., Wilensky, A.J., Friel, P.N., Levy, R.H. (1986) Ther. Drug Monitoring 8: 293–296.

19 Wilensky, A.J., Friel, P.N., Ojemann, L.M., Dodrill, C.B., McCormick, K.B., Levy, R.H. (1985) Epilepsia 6: 212–220.

20 Wagner, J.G., Sackellares, J.C., Donofrio, P.D., Berent, S., Sakmar, E. (1984) Ther. Drug Monitoring 6: 277–283.

21 Yagi, K., Seino, M., Mihara, T., Tottori, T., Numata, Y., Tsuji, M., Inoue, Y., Kudo, T., Watanabe, Y., Muranaka, H. (1987) Clin. Psychiatry 29: 111–119 (in Japanese).

22 Fukushima, K., Yagi, K., Seino, M., Fujiwara, T., Watanabe, M., Muranaka, H., Kudo, T., Okabe, M., Ohtani, K., Terauchi, N., Numata, Y. (1987) Jap. J. Pediatrics 40: 3389–3397 (in Japanese).

23 Matsumoto, K., Miyazaki, H., Fujii, T., Hashimoto, M. (1989) Chem. Pharm. Bull. 37: 1913–1915.

24 Loiseau, P., Schmidt, P., Deisenhammer, E., Despland, P.A., Egli, M., Bauer, G., Stenzel, E., Blankenhorn, V., Klinger, D., Turner, J., Wolf, R. (1987) 17th Epilepsy International Congress Abstracts p. 73.

25 Berent, S., Sackellares, J.C., Giordani, B., Wagner, J.G., Donofrio, P.D., Abou-Khalil, B. (1987) Epilepsia 28: 61–67.

26 Henry, T.R., Leppik, I.E., Gummit, R.J., Jacobs, M. (1988) Neurology 38: 928–931.

Recent Clinical Trials

New Antiepileptic Drugs (Epilepsy Res. Suppl. 3)
Editors: F. Pisani, E. Perucca, G. Avanzini, A. Richens
© 1991 Elsevier Science Publishers B.V. (Biomedical Division)

CHAPTER 24

Kinetics, metabolism and effects of carbamazepine-10,11-epoxide in man

Torbjörn Tomson[1], Leif Bertilsson[2]

[1]*Department of Neurology, Söder Hospital, Stockholm;* [2]*Department of Clinical Pharmacology at the Karolinska Institute, Huddinge Hospital, Huddinge, Sweden*

Carbamazepine is used in the treatment of epilepsy, trigeminal neuralgia and recently also in affective disorders[1]. After single oral doses of carbamazepine to healthy subjects a mean of 22% of

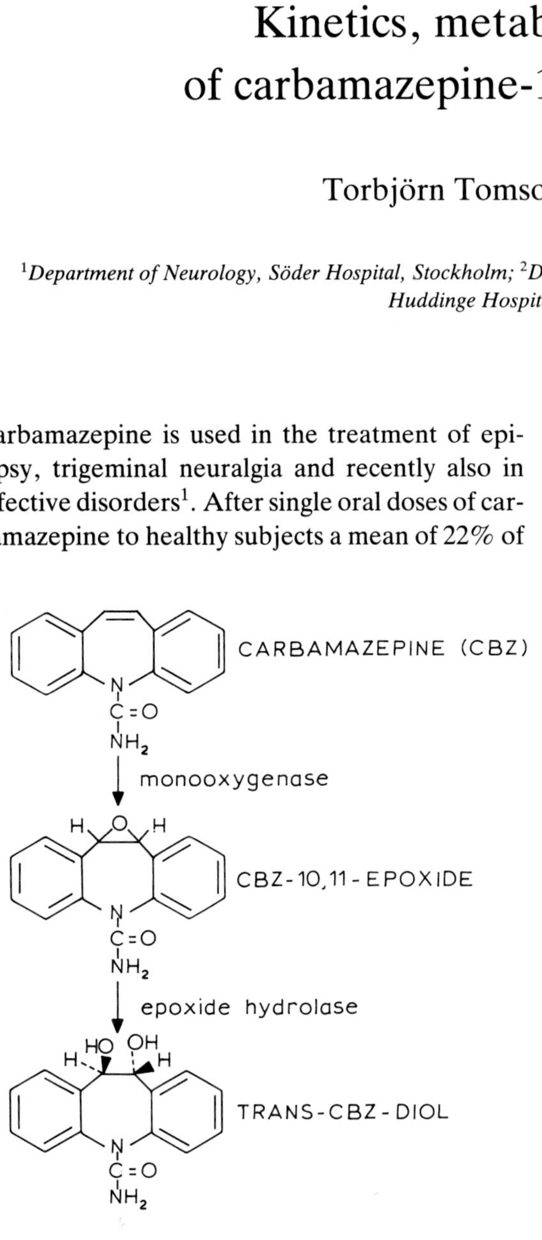

Fig. 1. Metabolism of carbamazepine by the epoxide-diol pathway to carbamazepine-10,11-epoxide and trans-carbamazepine-diol.

the dose was excreted as trans-10,11-dihydroxy-10,11-dihydrocarbamazepine (trans-carbamazepine-diol) (Fig. 1)[2]. The epoxide-diol metabolic pathway is induced during long term treatment with carbamazepine itself (autoinduction) and can be further induced by e.g. phenytoin and phenobarbitone (heteroinduction)[2]. The intermediate metabolite carbamazepine-10,11-epoxide has anticonvulsant properties similar to those of carbamazepine in different animal models[3].

Very early we developed an HPLC method for the simultaneous analysis of carbamazepine and carbamazepine-10,11-epoxide[4] and found that during carbamazepine treatment, the levels of the epoxide metabolite in plasma are about 20% of those of parent drug[1,3]. This percentage varies relatively little between and especially within patients.

The contribution of carbamazepine-10,11-epoxide to the overall effect of carbamazepine therapy has been debated for many years. It has been a common opinion that the epoxide to a large extent contributes to the side-effects of carbamazepine[1,3]. Most attempts to assess the effects of carbamazepine-10,11-epoxide have previously been made during administration of carbamazepine to patients. The results of such studies, how-

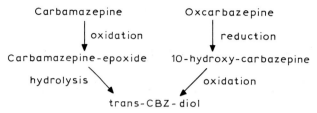

Fig. 2. Metabolism of carbamazepine and oxcarbazepine to the common end product trans-carbamazepine-diol.

ever, are ambiguous and it is clear that a reliable evaluation of the effects of the epoxide requires direct administration of the metabolite. This is the rationale behind the administration of carbamazepine-10,11-epoxide itself to healthy subjects and to patients. The aim of these studies was not to develop the epoxide to a new drug, but rather to collect novel information in order to improve the treatment with the parent drug carbamazepine.

We want to stress the difference between carbamazepine-10,11-epoxide and oxcarbazepine. Oxcarbazepine is a novel anticonvulsant, that is reduced to 10-hydroxy-carbazepine, which is hydroxylated to trans-carbamazepine-diol (Fig. 2). It has been claimed as an advantage that oxcarbazepine, in contrast to carbamazepine, is not metabolized to an epoxide. We have not found *any* evidence that there is a disadvantage to form carbamazepine-10,11-epoxide. In contrast we have not seen any side-effects in our clinical studies with this epoxide (see below).

Carbamazepine-10,11-epoxide is unstable in gastric juice and we administered the compound together with an antacid in the initial studies[5,6], but later an enteric coated tablet was used[7-9]. After a single oral dose of the epoxide the mean half-life is about 6 hours[5,7,8] (Fig. 3), which is much shorter than that of carbamazepine. As much as 90 ± 11 of the epoxide dose was recovered in urine as trans-carbamazepine-diol[5], which shows that carbamazepine-10,11-epoxide

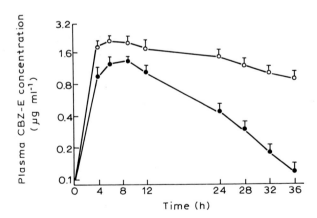

Fig. 3. Plasma concentrations of carbamazepine-10,11-epoxide (CBZ-E) (means±S.D.) in 6 healthy subjects after intake of a 100 mg enteric coated tablet of the epoxide before (●) and during (○) concurrent treatment with valpromide (300 mg b.i.d.). Reproduced by permission from Pisani et al.[8].

is metabolised mainly by epoxide hydrolase (Fig. 1). The plasma half-life of carbamazepine-10,11-epoxide was prolonged from 6.4 ± 1.4 to 20.5 ± 6.3 hours (p<0.01) during concomitant treatment with valpromide (Fig. 3; ref. 8). This is due to an inhibition of the epoxide hydrolase by valpromide[8]. Valproic acid is a less potent inhibitor of this enzyme[3]. In epileptic patients treated with phenytoin and/or phenobarbitone the half-life of the epoxide was shorter: 4.0 ± 0.8 hours (n=6), probably due to induction of epoxide hydrolase[3].

When carbamazepine-10,11-epoxide was given as single doses of 50 to 200 mg to healthy sub-

jects[3,5,7,8] peak plasma concentrations up to 16 µmol/l were seen. Side effects were noted in one study from Japan[10], in which 4 of 6 subjects reported dizziness and drowsiness, peculiarly first several hours after peak concentrations. None of the subjects in the other studies reported side effects.

The effects of carbamazepine-10,11-epoxide as maintenance therapy was evaluated in two pilot studies, in patients with trigeminal neuralgia[6] and with epilepsy[9]. Six patients with neuralgia had their optimal carbamazepine dose exchanged for carbamazepine-10,11-epoxide in suspension dur-

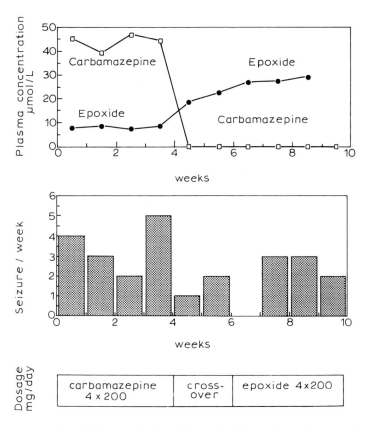

Fig. 4. The effect of carbamazepine-10,11-epoxide in epilepsy. Design and results of study as illustrated by one patient. Carbamazepine and carbamazepine-10,11-epoxide were dosed four times daily. Trough morning plasma levels of carbamazepine (open squares) and carbamazepine-10,11-epoxide (solid squares) are indicated in top panel. (Data from ref. 9).

TABLE 1

NUMBER OF SEIZURES IN 6 PATIENTS FIRST TREATED WITH CARBAMAZEPINE DURING 4 WEEKS AND THEREAFTER DURING 4 WEEKS WITH CARBAMAZEPINE-10,11-EPOXIDE IN EQUAL DOSES (DATA FROM REF. 9)

Patient	Treatment	
	Carbamazepine	Carbamazepine-10,11-epoxide
1	50	20
2	5	8
3	14	8
4	7	5
5	8	2
6	19	16

ing 3–6 days. Dosing was single blind. The pain control was essentially the same when carbamazepine and the epoxide were given in similar doses. No side effects were noted despite epoxide doses up to 1000 mg daily[6]. We could conclude that the epoxide had a more potent pain relieving effect than the parent drug on a plasma concentration basis. Seven patients were included in the study of carbamazepine-10,11-epoxide in epilepsy[9]. All had frequent complex partial seizures despite an optimal carbamazepine dose used in combination with another antiepileptic drug. Patients were studied during two 4-week periods, the first with the patients optimal carbamazepine dose, and the second with carbamazepine replaced, dose by dose, by carbamazepine-10,11-epoxide as enteric coated tablets. Dosing was single blind and the drugs given q.i.d. One patient withdrew from the trial after generalized seizures in the washout week following cross-over from carbamazepine to the epoxide. The design and results of the study are illustrated in Fig. 4.

Mean steady-state concentrations during carbamazepine treatment were 32 µmol/l for parent drug and 7.3 µmol/l for the epoxide metabolite. The mean carbamazepine-10,11-epoxide concentration was 25 µmol/l during treatment with the epoxide. However, there were large and unpredictable diurnal variation in plasma epoxide levels during epoxide therapy (Fig. 5). These fluctuations hampered the assessment of the therapeutic effects of the epoxide. However, no significant difference in seizure frequency was found between the carbamazepine and the epoxide periods (Table 1). Neuropsychological evaluation revealed a significant improvement in finger motor speed and logical reasoning during epoxide therapy[9].

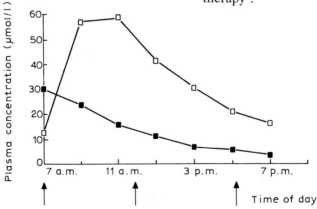

Fig. 5. Plasma carbamazepine-10,11-epoxide concentration profiles, 7 a.m. to 7 p.m. in two patients treated with carbamazepine, 300 mg q.i.d. Time of dosing indicated by arrows. (Data from ref. 9).

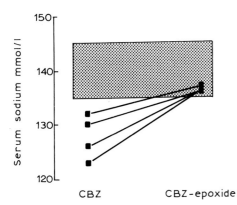

Fig. 6. Four patients who developed hyponatremia during carbamazepine treatment. Serum sodium levels before and three weeks after changing from carbamazepine-10,11-epoxide in equal doses.

Sodium levels in serum were significantly higher when patients were treated with the epoxide[9]. To explore this observation further, we selected four patients, who had developed hyponatremia during carbamazepine treatment. Carbamazepine was exchanged, dose by dose, for carbamazepine-10,11-epoxide during three weeks. Sodium levels in serum were normalized in all (Fig. 6), indicating that the epoxide does not share the antidiuretic properties of carbamazepine.

In conclusion, it is unlikely that carbamazepine-10,11-epoxide is a major contributor to the side-effects of carbamazepine. In some respects the metabolite appears to be less toxic than the parent drug. Further studies are needed to establish the antiepileptic potential of this compound. To this end, a slow release formulation needs to be developed.

References

1 Bertilsson, L., Tomson, T. (1986) Clin. Pharmacokin. 11: 177–198.
2 Eichelbaum, M., Tomson, T., Tybring, T., Bertilsson, L. (1985) Clin. Pharmacokin. 10: 80–90.
3 Kerr, B.M., Levy, R.H. (1989) In: Levy, R.H., Mattson, R., Meldrum, B., Penry, J.K., Dreifuss, F.E. (eds.), Antiepileptic Drugs, Third Edition, Raven Press, New York, pp. 505–520.
4 Eichelbaum, M., Bertilsson, L. (1975) J. Chromatogr. 103: 135–140.
5 Tomson, T., Tybring, G., Bertilsson, L. (1983) Clin. Pharmacol. Ther. 33: 58–65.
6 Tomson, T., Bertilsson, L. (1984) Arch. Neurol. 41: 598–601.
7 Spina, E., Tomson, T., Svensson, J.O., Bertilsson, L., Faigle, J.W. (1988) Ther. Drug. Monit. 10: 382–385.
8 Pisani, F., Fazio, A., Oteri, G., Spina, E., Perucca, E., Bertilsson, L. (1988) Br. J. Clin. Pharmacol. 25: 611–613.
9 Tomson, T., Almkvist, O., Nilsson, B.Y., Svensson, J.O., Bertilsson, L. (1990) Arch. Neurol. In press.
10 Sumi, M., Watari, N., Umezawa, O., Kaneniwa, N. (1987) J. Pharmacobiodyn. 10: 652–661.

New Antiepileptic Drugs (Epilepsy Res. Suppl. 3)
Editors: F. Pisani, E. Perucca, G. Avanzini, A. Richens

CHAPTER 25

Gabapentin

David Chadwick

Walton Hospital, Liverpool, UK

Gabapentin (1-(aminomethyl) cyclohexaneacetic acid) is a GABA-analogue that is well absorbed and penetrates the blood-brain barrier. It is effective against a variety of seizures in animals, particularly those caused by interference with GABAergic transmission or provoked by excitatory amino acids[1]. In spite of this the precise mechanism of its antiepileptic properties is unclear as gabapentin does not affect whole brain levels of GABA or have significant effects on GABA transaminase activity. It does not show any significant binding to GABA or benzodiazepine receptors and does not influence neuronal uptake of GABA at pharmacologically relevant doses. Our experience with the clinical use of this drug will be described here.

A double-blind dose ranging three-way cross-over study of gabapentin as add-on therapy[2]

This first study was designed to compare the efficacy of three different doses of gabapentin in 25 patients with severe drug-resistant partial and generalised epilepsies. Each treatment period was of 2 months duration. The results indicated that 900 mg/day of gabapentin had a greater antiepileptic effect than 300 and 600 mg/day against all seizure types considered together. There was evidence of a dose related antiepileptic effect that was statistically significant when all seizures were compared together and similar trends were seen

for partial seizures and tonic-clonic seizures separately (Fig. 1). The median frequency of all seizures was reduced by 30% compared to baseline (a 50% reduction for partial seizures and a 70% reduction for tonic-clonic seizures). Whilst the direct dose to dose comparisons inferred antiepileptic efficacy, this study could not be regarded as providing definite proof of efficacy. There was no random allocation between baseline period and drug dosage and placebo was not given during the baseline period. Whilst it seemed highly unlikely that the results could be explained by a placebo effect this trial design could not exclude such a possibility.

The study showed little evidence of adverse effects and detailed psychometric testing failed to show any significant changes at the three different doses from baseline. Importantly, there was no evidence that gabapentin had significant interactions with other antiepileptic drugs. This was expected as the drug is excreted unmetabolised and does not undergo significant protein binding.

A double-blind, placebo-controlled, parallel group study of 1200 mg/day gabapentin as add-on therapy in partial epilepsy[3]

This study was undertaken by a group of collaborators on a multicentre basis.

Patients with partial epilepsy resistant to treatment with one or two antiepileptic drugs were fol-

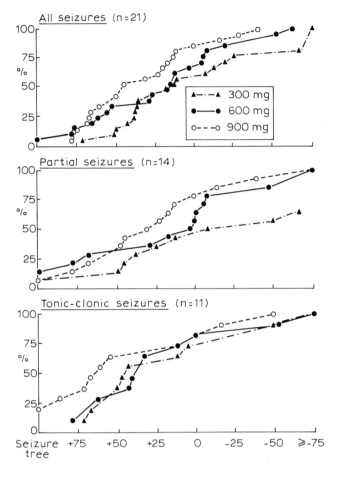

All seizures (n=21)

Partial seizures (n=14)

Tonic-clonic seizures (n=11)

▲–·–▲ 300 mg
●——● 600 mg
o– – –o 900 mg

Seizure +75 +50 +25 0 -25 -50 ≥-75
tree

Fig. 1. Cumulative distribution of patients as their percentage change in seizure frequency from baseline at doses of 300 mg, 600 mg and 900 mg gabapentin, for all seizures, partial seizures and tonic-clonic seizures. Vertical axis is % of N patients showing appropriate change in seizure frequency or better. Horizontal axis is change in seizure frequency relative to baseline period (O = no change, positive values = % improvement, negative values = % deterioration.
Differences as follows:

| | | Wilcoxon-Wilcox | |
	Friedman	900 vs 300	900 vs 600
All seizures	$p < 0.001$	$p = 0.01$	$p = 0.05$
Partial	$p = 0.034$	$p = 0.05$	$p = n.s.$
Tonic-clonic	$p = 0.044$	$p = n.s.$	$p = n.s.$

lowed for three months to establish a baseline seizure frequency. Sixty one were randomized to receive gabapentin, 66 to receive placebo. There were no significant differences between the baseline variables for the two treatment groups (Table 1). Patients were evaluated for a 12 week period after an initial 2 week titration period.

Data from 14 patients were not evaluable for efficacy. The frequency of partial seizures was at least halved in 25% of patients treated with gabapentin compared to 9.8% treated with placebo (P = 0.043, Fisher's Exact Test). On an intention to treat analysis 23% of patients randomised to

gabapentin were responders by at least a 50% reduction in fit frequency compared with 9% of patients who received placebo (P = 0.049).

The median percentage change in partial seizure frequency from baseline was greater with gabapentin (−29.2%) than placebo (−12.5%). In both gabapentin and placebo groups, the median reduction in seizure frequency was greater for patients in whom partial seizures did not show secondary generalisation. The reduction in seizure frequency was apparent by the end of the 2 week titration phase and persisted through the full 3 months treatment. No significant changes in

TABLE 1

PATIENT CHARACTERISTICS AND REASONS FOR EXCLUSION

	Gabapentin (n = 61)	Placebo (n = 66)
Age (yr)[*]	30 (15–62)	31 (14–73)
Sex (M.F.)	39%, 61%	44%, 56%
Duration of epilepsy (yr)#	17 (2–47)	19 (4–38)
Standard antiepileptic drugs taken		
0	1	0
1	20	21
2	38	43
3	2	2
Seizures/28 days during baseline period		
All seizures#	13 (3–368)	13 (1–216)
Secondary tonic-clonic#	5 (0.3–47)	4 (0.3–32)
Reasons for exclusion		
<1 seizure per week baseline	1	2
<56 days seizure diary	6	2
Study drug stopped >14 days	3	1
Total exclusions	9	5

[*]Mean (range).
#Median (range).

plasma concentrations of other anticonvulsants were observed during the study. When response ratio (a measure of change in seizure frequency[3]) was plotted against gabapentin plasma concentrations the negative slope was significantly different from zero and indicated an increased efficacy with increased gabapentin plasma concentrations.

Thirty-eight of 61 patients (62%) who received gabapentin and 27 of 66 (41%) who received placebo reported an adverse event during the study. The most frequent reports in the gabapentin group were somnolence (14.8%), fatigue (13.1%), dizziness (6.6%), and increased weight (4.9%). In the placebo group, headache (9.1%) was the most commonly reported symptom, followed by dizziness (4.5%) and somnolence (4.5%). Most such symptoms were rated as mild

to moderate, not disabling, and, in the gabapentin group, resolved during the study. Eleven of 127 patients withdrew from the study because of adverse events. Seven patients who received gabapentin withdrew because of: tiredness and poor memory; low white cell count in a patient also treated with carbamazepine; post-ictal speech disturbance and unilateral weakness (not previously experienced); vomiting and drowsiness; jitteriness; increased seizure frequency; and a mild maculopapular rash. Four patients withdrew during treatment with placebo because of: altered mental state; confusion, odd behaviour and paranoia; generalised seizures; and a general feeling of ill-health. Haematological and biochemical parameters monitored during treatment showed no significant trends for any parameter compared to baseline for either gabapentin or placebo.

Long-term gabapentin treatment

Following completion of this double-blind cross-over study of 1200 mg/day gabapentin[3] as add-on therapy patients who had received gabapentin in the blinded phase and who had benefited could elect to continue therapy whilst those who received placebo could commence open-label add-on gabapentin (1200 mg/day).

Thirty-one patients who had received double-blind gabapentin elected to continue open-label gabapentin, of whom 27 continued for a further 3 and 6 months and 21 continued for a further 9 months. Compared to baseline, median percentage seizure reduction was −33% for the 3–6 month period, −38% for 6–9 months, and −41% for 9–12 months. Six patients discontinued gabapentin during open-label because of loss of efficacy and 3 because of adverse events possibly (diplopia, and increased seizure frequency) and probably not (breast carcinoma) related to gabapentin.

Fifty-seven patients who received placebo during double-blind elected to receive open-label gabapentin. Forty five completed 3 months' therapy, 33 completed 6 months' therapy, 25 completed 9 months' therapy, and 19 completed 12 months' therapy. After 3 months median seizure reduction was 41%; 43% after 6 months, 20% after 9 months and 40% after 12 months. Twelve patients stopped treatment because of lack of efficacy, 9 because of adverse events possibly related to treatment.

Conclusions

These studies indicate a significant antiepileptic effect for gabapentin which may be comparable with that for conventional antiepileptic drugs and which appears to be maintained over 12 months. Adverse events with this drug are relatively few and relatively mild. The drug may therefore be a valuable introduction in potentially being better tolerated than some existing antiepileptic drugs.

References

1 Schmidt, B. (1989) In: Levy, R., Dreifuss, F.E., Mattson, R.H., Meldrum, B.S., Penry, J.K. (eds.), Antiepileptic Drugs, Raven Press, New York, 925–935.
2 Crawford, P., Ghadiali, E., Lane, R., Blumhardt, L., Chadwick, D. (1987) J. Neurol. Neurosurg. Psychiatry 50: 682–686.
3 U.K. Gabapentin Study Group (1990) Lancet (in press).

New Antiepileptic Drugs (Epilepsy Res. Suppl. 3)
Editors: F. Pisani, E. Perucca, G. Avanzini, A. Richens
© 1991 Elsevier Science Publishers B.V. (Biomedical Division)

CHAPTER 26

Lamotrigine in refractory epilepsy: a long-term open study

Francesco Pisani, Maria Russo, Rosa Trio, Caterina Artesi, Antonino Fazio, Giancarla Oteri, Raoul Di Perri

First Neurological Clinic of the University of Messina, 98013 Contesse-Messina, Italy

Introduction

The antiepileptic action of lamotrigine in man has been so far evaluated in four controlled[1-4] and approximately thirty open studies[5]. Furthermore, a recent observation suggests that the drug may be effective in status epilepticus[6].

We report here the results of long-term treatment with lamotrigine in a group of patients with severe epilepsy with the main aim of assessing the risk/benefit ratio of this drug during maintenance therapy.

Patients and Methods

Patient selection

Patients were admitted into the study on the basis of the following inclusion criteria: age 12–70 years; male or female of non-childbearing potential; diagnosis of epilepsy uncomplicated by suspected pseudo-seizures and uncontrolled by conventional treatment, irrespective of seizure type; no episodes of status epilepticus in the past 6 months; at least two fits per month in the previous 3 months; antiepileptic medication (no more than 3 antiepileptic drugs) stable for the preceding 3 months; no regular drugs other than anticonvulsants or oral contraceptives; no

hepatic, renal, cardiac, progressive neurological disease or major psychiatric disorder; no abnormal laboratory values of clinical significance; hepatitis B surface antigen negative; no alcohol or drug abuse; no past record of poor compliance, and written consent.

Drug dosage

Lamotrigine was added to pre-existing therapy at an initial daily dose of 100–200 mg given in two equally divided administrations and adjusted up to a maximum of 600 mg according to clinical response. The dosage of associated anticonvulsants was maintained unchanged for the first 12 months. Following this period, patients with a satisfactory response to lamotrigine went directly into the long-term phase of the study. Response to lamotrigine was considered satisfactory when at least a 50% reduction in seizure frequency or a clinically important improvement in quality of life due to decreased seizure duration or severity occurred. During the long-term phase lamotrigine dosage and regimen were maintained unchanged. Withdrawal of associated drugs was attempted in those patients who gave their informed consent. Phenytoin and carbamazepine were gradually discontinued within 1–2 months and phenobarbitone within 2–3 months.

TABLE 1

DETAILS OF THE PATIENTS ADMITTED TO THE TRIAL, DRUG THERAPY AND TOTAL SEIZURE FREQUENCY

Pt	Sex	Age (yrs)	B.W. (kg)	Seizure type	Comedication (mg/day)	Months on LTG	LTG dose (mg/kg)	Plasma LTG (μg/ml)§	Total seizure frequency per month	
									baseline	last 3 months on LTG
GF	F	45	41	CP+SG	CBZ1200	12	200	1.67	13.3	14.6
IS	M	25	65	CP+SG	PB150	23	300	2.79	32.0	27.0
LS	M	49	86	CP	CBZ1000	1	400	1.50	36.0	early drop-out
ZV	M	52	85	CP+SG	CBZ1200	23	400	4.03	12.0	8.0
CO	F	30	48	CP	CBZ600+PHT325	12	150	1.56	15.0	12.0
CA	F	26	60	CP+SG	CBZ1200	7days	200	/	69.0	early drop-out
VL	F	16	52	CP	CBZ1200+PHT300	15days	400	1.32	25.0	early drop-out
CG	M	22	70	CP+SG+ AT+AA	CBZ1200+PB100	12	600	3.65	40.0	31.3
PD	F	18	65	CPSG	PB200+CBZ800*	23	500	7.50	8.3	6.3 (6.3)$
BA	M	33	80	CP	CBZ1000	5days	200	/	23.0	early drop-out
PR	F	37	70	CP+SG	CBZ1000+PB200	23	300	4.00	109.0	55.0
DG	M	43	63	CP+SG	CBZ1400+PB150	7days	200	/	51.0	early drop-out
GE	M	46	50	CP	CBZ1400+PRM1000	23	400	4.60	122.3	35.5
SE	M	34	80	CP+SG	CBZ1200+PB200*	23	400	3.24	49.6	26.3 (29.0)$
AM	M	17	58	CP+SG	CBZ1200	12	400	2.23	21.6	16.6
MC	M	45	68	CP	PHT600	12	400	2.98	20.3	16.3
PP	F	29	64	CP+SG	CBZ1200+PB200*	23	400	2.04	33.3	9.3 (9.0)$
PF	M	32	68	CP+SG	CBZ1200+PB200	23	400	2.84	44.0	18.0
DM	M	15	43	CP	CBZ1200	23	200	1.26	5.0	3.0
PT	F	60	57	CP	CBZ2000+PHT450	23	450	6.84	8.0	4.3
RM	F	14	30	CP	CBZ1400+PB200	11	250	1.96	11.0	11.6
AS	F	12	44	CP+SG+ AT+AA	ESM1250+VPA800 +PB150*	23	350	2.93	66.0	38.0 (40.3)$
LGA	M	30	72	CP	PHT600*	23	400	2.28	6.6	3.0 (3.0)$
QC	F	15	34	CP+SG	CBZ1000	23	100	1.01	21.0	9.0
IG	F	29	70	CP+SG	PHT250	12	400	3.34	7.0	17.6
SC	M	36	73	CP	PB100	11	400	2.51	6.3	5.0
AA	F	18	72	CP+SG+ AT+AA	CBZ1000+PB200	23	400	1.03	84.3	20.6
DA	M	16	70	CP+SG	CBZ1000+PB100	23	400	0.75	10.3	4.0
SS	M	23	70	CP	CBZ1400+PHT350	23	200	2.23	19.6	9.0
BC	F	38	50	CP	CBZ600+PB100	10	150	6.35	3.0	13.3
BD	F	29	65	CP	CBZ1000	11	400	2.79	8.0	7.0
CR	F	30	75	CP	CBZ600	23	300	3.86	4.3	3.0

§= mean of 3 determinations in the last 3 months of LTG therapy; *= drug discontinued after 1 year of LTG therapy; $= total seizure frequency per month in the last 3 months before discontinuation of associated drug*; AA= atypical absences; AT= atonic seizures; CBZ= carbamazepine; CP= complex partial seizures; ESM= ethosuximide; LTG= lamotrigine; PB= phenobarbitone; PHT= phenytoin; SG= secondarily generalized seizures; VPA= valproic acid.

Safety and efficacy measurements

Each patient was seen regularly in the clinic at one-month intervals over the first year and at 2-month intervals thereafter.

Adverse events were recorded and classified according to the reported intensity (mild, moderate or severe) and seriousness (non-serious, serious, life-threatening). Vital signs, haematology, biochemistry, urinalysis and plasma drug levels were assessed at regular intervals.

Seizure frequency was recorded by the patients on a fit calendar. The number of total (i.e. all seizure types), partial (simple or complex) and secondarily generalized seizures was compared to that of the three months preceding lamotrigine treatment. Seizures were classified according to the international classification.

Plasma concentrations of conventional antiepileptic drugs were measured by EMIT. Lamotrigine was determined by a specific HPLC method[7].

Results

Thirty-two patients fulfilled the inclusion criteria and were admitted to the trial (Table 1).

Drop outs and adverse events

Five patients dropped out, 4 within the first 2 weeks because of a cutaneous rash and 1 after one month because of development of hallucinations. Dizziness (n=4), headache (n=3), drowsiness (n=1), hand tremor (n=1) and ataxia (n=1) were the most common side effects during the first days of lamotrigine treatment.

These effects were usually mild or moderate and disappeared with time. An increase in body weight was observed in 3 patients. Two patients had their first generalized tonic-clonic seizure while on lamotrigine treatment (BC during the 3rd month and CO during the 7th month) (Table 1).

Laboratory evaluation

No changes in safety laboratory parameters were observed.

Efficacy data

In the 27 patients who remained in the trial, the median monthly total seizure frequency decreased from 17 (range 3–122) at baseline to 12 (range 3–55) during the last 3 months on lamotrigine (p< 0.001, Wilcoxon rank-sum test) (Table 1 and Fig. 1).

Ten out of the 27 patients discontinued lamotrigine within the first 12 months because of lack of efficacy, while 17 continue on the trial with a follow-period of 23 months (Fig. 1). In this group, 11 patients showed a >50% reduction of seizure frequency and 6 showed a reduction of seizure duration and intensity. A summary of the data obtained in these patients is given in Table 2.

A strong response to lamotrigine in both atypical absences and atonic seizures (reduction >80%) was observed in the only 3 patients suffering from these types of seizures.

Drug withdrawal

Discontinuation of one of the associated antiepileptic drugs without deterioration of the clinical improvement achieved with lamotrigine was possible in 5 patients. One of these patients is now on lamotrigine monotherapy after phenytoin withdrawal. In the 5 patients who underwent drug withdrawal fit frequency remained virtually unchanged as compared with the pre-withdrawal period (Table 1).

Plasma anticonvulsant drug levels

During the baseline period plasma drug levels (μg/ml, mean±SD) were 7.9 ± 2.7 for carbamazepine (n=21), 19.1 ± 6.2 for phenytoin (n=6), 31.7 ± 10.7 for phenobarbitone (n=14), 65.0 for ethosuximide (n=1) and 70.9 for valproic acid (n=1). These values remained virtually unchanged during lamotrigine treatment. In 8 patients on carbamazepine-lamotrigine co-medi-

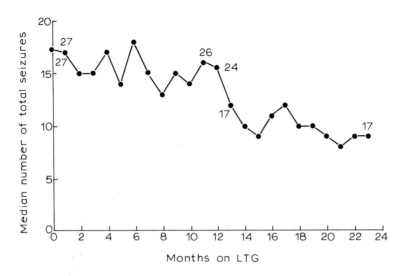

Fig. 1. Median monthly total seizure frequency during the baseline period (time 0) and during lamotrigine therapy in the 27 patients who received lamotrigine for more than one month. Numbers indicate the number of patients in each month. LTG= lamotrigine.

cation the plasma levels of carbamazepine-10,11-epoxide (1.19 ± 0.6 µg/ml) were not statistically different from those observed in a control group of 8 patients on carbamazepine monotherapy (0.97 ± 0.22 µg/ml, p >0.05, Student's t test for unpaired data).

Plasma lamotrigine levels ranged from 1.01 to 7.50 µg/ml and did not show any apparent correlation with clinical response.

Discussion

Our observations suggest that lamotrigine is generally well tolerated during maintenance therapy. Side effects were usually mild, disappeared with time and were apparently unrelated to plasma lamotrigine concentrations. In fact, in some patients (PD, PT and BC), abnormally high levels of lamotrigine did not cause any apparent toxic symptom. The only relatively frequent adverse effect which induced withdrawal of the drug

TABLE 2

MEDIAN SEIZURE FREQUENCY FOR TOTAL, PARTIAL AND SECONDARILY GENERALIZED SEIZURES IN THE 17 PATIENTS ON LONG-TERM LAMOTRIGINE THERAPY

Seizure type	Median seizure frequency		Per cent reduction
	Lamotrigine[*]	Baseline	
Total	25 (range: 5–150)	69 (range: 12–366)	63.7
CP	22 (range: 5–103)	60 (range: 9–366)	63.3
SG	2 (range: 0–100)	18 (range: 9–180)	88.8

[*]Last 3 months of lamotrigine treatment. Baseline (by history) = the 3 months preceding LTG therapy. CP = complex partial seizures; SG = secondarily generalized seizures.

was a skin rash, which occurred in 12.5% of patients.

With regard to efficacy, the present trial suggests that lamotrigine is effective against both partial and generalized seizures, in agreement with the results of previous controlled studies[8]. Further investigations should clarify whether lamotrigine is also active against other types of seizure, such as atonic fits and atypical absences, as our observations and those of other authors[9,10] seem to suggest.

References

1 Oxley, J., Sander, J. (1988) Wellcome Research Laboratories, internal report.
2 Jawad, S., Richens, A., Goodwin, G., Yuen, W.C. (1989) Epilepsia 30: 356–363.
3 Binnie, C.D., Debets, R.M.C., Engelsman, M., Meijer, J.W.A., Meinardi, H., Overweg, J., Peck, A.W., Van Wieringer, A., Yuen, C. (1989) Epilepsy Res. 4: 222–229.
4 Loiseau, P., Bes, A., Menager, T., Duche, B., Arne-Bes, M.C. (1990) 7a Riunione Congiunta delle Leghe Italiana, Francese, Portoghese e Spagnola contro l'Epilessia, Abstracts Book, p. 78.
5 Betts, T. (1990) Wellcome Research Laboratories, internal report.
6 Pisani, F., Gallitto, G., Di Perri, R. (1991) J. Neurol. Neurosurg. Psychiatry (in press).
7 Cohen, A.F., Land, G.S., Breimer, D.D., Yuen, W.C., Winton, C., Peck, A.W. (1987) Clin. Pharmacol. Ther. 42: 202–208.
8 Richens, A., Yuen, A.W.C. (1989) 18th International Epilepsy Congress, Abstract Book, p. 142.
9 Betts, T., Pigott, C., Grace, E. (1989) 18th International Epilepsy Congress, Abstract Book, p. 143.
10 Wallace, S.J. (1989) 18th International Epilepsy Congress, Abstract Book, p 143.

New Antiepileptic Drugs (Epilepsy Res. Suppl. 3)
Editors: F. Pisani, E. Perucca, G. Avanzini, A. Richens
© 1991 Elsevier Science Publishers B.V. (Biomedical Division)

CHAPTER 27

Vigabatrin

Roberto Michelucci, Carlo Alberto Tassinari

Neurological Clinic, University of Bologna, School of Medicine, Bologna, Italy

Introduction

Among drugs designed to modify GABA activity, vigabatrin (γ-vinyl GABA) has proved to be a highly specific irreversible inhibitor of GABA-transaminase, the enzyme that metabolizes GABA, and to raise GABA concentrations in the central nervous system of laboratory animals and man[1]. In a variety of animal seizure models, these increases in brain GABA levels have been associated with anticonvulsant activity[2]. Open and placebo-controlled clinical studies, in which vigabatrin was added to preexisting anticonvulsant regimens have demonstrated reductions in seizure frequency in most subjects with good tolerance[1,3–9].

In 1983 a clinical study of vigabatrin for refractory epilepsy was initiated in our Institute. Since then 51 patients have received the drug under controlled conditions and our experience now includes 2 short-term studies and 1 long-term investigation[9–10]. The results of these studies will be summarized in this paper.

Materials, Methods and Results

Double-blind study[9]

Thirty-one patients (15 female and 16 male), aged 10 to 58 years, with at least four documented seizures per month while receiving optimal doses of current antiepileptic medication, entered the study. Thirty patients suffered from partial epilepsy whereas one patient had progressive myoclonus epilepsy. All patients presented with severe cases of epilepsy, as demonstrated by the high mean ($\pm$ SD) weekly seizure frequency (12.2 $\pm$ 17.8) during the baseline period.

The study consisted of 4 treatment periods covering a total of 9 months. After a 2 month run-in period, each patient received placebo or vigabatrin (2 to 3 g/d, stratified according to weight) for 3 months. This was followed by an abrupt crossover to the alternative treatment (vigabatrin or placebo) that was also administered for 3 months. At the end of the double-blind treatment periods, all patients received placebo for one month under single-blind conditions. Seizure type and frequency were noted in diaries, as were any adverse events experienced during the study. Monitoring at each clinic visit included physical and neurological examination, plasma concentration of concomitant antiepileptic medications, electroencephalograms (EEGs) and hematological and biochemical tests. Statistical analyses of seizure frequency and plasma concentrations of concomitant drugs were performed using the method outlined by Hills and Armitage[11].

One patient was withdrawn from the study because of the development of leukopenia, thought

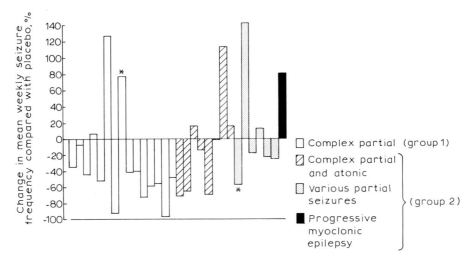

Fig. 1. Percent change in mean weekly seizure frequency during vigabatrin period compared with placebo (from Tassinari et al., 1987, with permission). For definition of Group 1 and Group 2, see text.

to be due to concomitant treatment with trime-thadione. Thirty patients completed the study. Statistically, no significant difference was observed between the mean weekly seizure frequencies of the vigabatrin and placebo periods. Despite this, when the patient population was divided into 2 diagnostic categories, a differential response to vigabatrin of the 2 subgroups was documented on the examination of frequency data (Fig. 1). In the 15 patients (group 1) presenting with only a single seizure type, unifocal EEG abnormalities, a mean weekly seizure frequency lower than 5 and absence of mental impairment, there was a statistically significant decrease in seizure frequency (P <.02) during the vigabatrin period. On the other hand, no significant treatment effect was found for the remaining 15 patients, who presented with various partial seizures, multifocal EEG abnormalities, a mean weekly seizure frequency higher than 10 and intellectual problems.

Unwanted effects were usually mild and transitory, with drowsiness being the most frequently reported side effect. No clinically significant

changes in safety tests were observed during treatment with vigabatrin. The 19 patients taking phenytoin showed a significant decrease (P <.01) of plasma levels of this drug during the vigabatrin period.

Single-blind study

Twenty patients (13 female and 7 male), aged 15 to 36 years, with drug-resistant epilepsy entered the study. Nineteen patients suffered from partial epilepsy whereas one patient had progressive myoclonus epilepsy.

The study consisted of 3 treatment phases, covering 7 to 11 months. Following a 3-month baseline period (phase I) (which included a one-month 'no additional treatment' period and a 2-month placebo period), each patient received add-on vigabatrin at a dose of 2 g/d for 2 months (phase II). This was followed by a 6-month dose modification phase, to titrate the optimal dose for each patient (phase III). The daily dosage of vigabatrin was increased or decreased by 0.5–1 g, if clinically indicated, at 2-month intervals. Monitoring was the same as described in the double-blind study. Evaluation of efficacy was based on monthly seizure frequency calculated from patient's calendars. Efficacy ratings performed at the end of phase II and phase III were compared with baseline seizure frequency by a Wilcoxon's rank test.

All patients completed phase II of the trial. Four of them showed a slight increase in seizure frequency or experienced side effects (drowsiness, confusion) and were consequently withdrawn at the end of phase II. The remaining 16 patients completed the study. Twelve of them (75%) had a greater than 50% reduction in seizure frequency, whereas 2 patients had their frequency reduced by 40–50%. Two patients were unchanged. Between baseline and dose modification phase, the total frequency of seizures decreased significantly (from 59.4 ± 58.8 to 28.8 ± 45.8) (P <.02). The mean optimal dose at the end of the study was 3 g/d. The 7 patients taking phenytoin had a significant decrease of plasma levels of this drug during the vigabatrin period (P <.01). The safety tests were unchanged.

Long-term study

Twenty-four drug-resistant epileptic patients (13 male and 11 female), aged 18 to 48 years, who had been significantly improved by vigabatrin during the previous 2 short-term placebo-controlled trials, entered the long-term follow-up study. Of these, 23 had partial epilepsy whereas 1 patient suffered from progressive myoclonus epilepsy. All patients were treated on an outpatient basis, with regular visits at 3 month intervals. Seizure calendars were recorded at each visit, along with results of safety tests (see double-blind study). Statistical analysis of seizure frequency was performed using the paired Wilcoxon's rank test.

Seven patients were withdrawn from the study (5 because of insufficient efficacy and 2 for reasons unrelated to treatment) after having re-

ceived vigabatrin for 2 to 24 months (median, 12 months). The mean duration of therapy in the 17 patients who continued vigabatrin was 37.4 ± 7.9 months. During the first 6 months on vigabatrin treatment, the mean seizure frequency dropped to less than 50% compared with placebo and showed no tendency to increase with time. In all patients (including the drop-outs) the mean monthly seizure frequency was 41.6 ± 37.6 on placebo and 21.4 ± 22.3 during the last 2 months on vigabatrin (P <.01). The most frequently reported side effect was weight gain. The safety tests were unchanged. A significant decrease in plasma phenytoin levels was observed during vigabatrin treatment. Vigabatrin was given at a mean dosage of 49.4 ± 12.5 mg kg^{-1} body weight during the last 2 months of therapy.

Discussion and conclusions

Our single and double-blind trials were associated with an improvement (seizure reduction by more than 50% as compared to baseline) in 60% and 33% of patients respectively. The lower efficacy observed in the double-blind study was due, in our opinion, to a wider electro-clinical spectrum and heterogeneity of the patients included[9]. Vigabatrin retained anticonvulsant efficacy during long-term treatment, with 70% of patients being still significantly improved after a mean period of 37 months. Other studies of vigabatrin for refractory epilepsy report results similar to ours[3–8,12].

Mild and transitory drowsiness was the most frequently reported side effect during short-term studies and weight gain was the only effect observed during long-term treatment. Indeed, other clinical trials to date indicate no significant toxicity associated with short- or long-term vigabatrin administration[3–8,12].

We constantly observed a marked decrease in phenytoin plasma levels during vigabatrin treatment. The mechanism of this interaction, however, has not been elucidated[13].

In conclusion, the combination of high response rate and good tolerability make vigabatrin a particularly promising therapy for refractory partial seizures.

References

1 Schechter, P.J., Hanke, N.F.J., Grove, J., Huebert, N., Sjoerdsma, A. (1984) Neurology 34: 182–186.
2 Kendall, D.A., Fox, D.A., Enna, S.J. (1981) Neuropharmacology 20: 351–355.
3 Rimmer, E.M., Richens, A. (1984) Lancet 1: 189–190.
4 Gram, L., Klosterskov, P., Dam, M. (1985) Ann. Neurol. 17: 262–266.
5 Loiseau, P., Hardenberg, J.P., Pestre, M., Guyot, M., Schechter, P.J., Tell, G.P. (1986) Epilepsia 27: 115–120.
6 Tartara, A., Manni, R., Galimberti, C.A., Hardenberg, J., Orwin, J., Perucca, E. (1986) Epilepsia 27: 717–723.
7 Browne, T.R., Mattson, R.H., Penry, J.K., Smith, D.B., Treiman, D.M., Wilder, B.J., Ben-Menachem, E., Napoliello, M.J. Sherry, K.M., Szabo, G.K. (1987) Neurology 37: 184–189.
8 Cocito, L., Maffini, M., Perfumo, P., Roncallo, F., Loeb, C. (1989) Epilepsy Res. 3: 160–166.
9 Tassinari, C.A., Michelucci, R., Ambrosetto, G., Salvi, F. (1987) Arch Neurol 44: 907–910.
10 Michelucci, R., Tassinari, C.A. (1989) Br. J. Clin. Pharmac. 27: 119S–124S.
11 Hills, M., Armitage, P. (1979) Br. J. Clin. Pharmac. 8: 7–20.
12 Tartara, A., Manni, R., Galimberti, C.A., Mumford, J.P., Iudice, A., Perucca, E. (1989) J. Neurol. Neurosurg. Psychiatry 52: 467–471.
13 Rimmer, E.M., Richens, A. (1989) Br. J. Clin. Pharmac. 27: 27S–33S.